THE PARALLEL BANG

The explosive growth of human understanding in the 21st century

Jack Bacon

Normandy House Publishers

With a foreword by
Paul Davies

Copyright© 2008 Jack Bacon
All rights reserved. With the exception of cited quotations of two paragraphs or less, and for excerpts used in reviews of this book, no part of this book may be reproduced or transmitted in any form or by any means, electronic or mechanical, including photocopying, recording, web posting, or any information storage and retrieval system, without permission from the publisher.

First published in the United States of America in 2006 by Normandy House Publishers.

Publisher's Cataloging-in-Publication

Bacon, Jack 1954-
 The Parallel Bang: The explosive growth of human understanding in the 21st century/ Jack Bacon
 1st ed.
 p. cm.
 Includes bibliographical references and index.
 LCCN:
 ISBN: 0-9708319-3-5

 1. Technology—Social Aspects. 2. Technological Innovations.
 3. Cyberspace. I. Title.

T14.5.B132 2006   303.483

Copies of this and other Normandy House Publishers products can be ordered from:
  Normandy House Publishers
  P.O. Box 59-1066
  Houston, TX 77259-1066
www.normandyhousepublishers.com
  Printed in USA
10 9 8 7 6 5 4 3

Fidelité Fecundité Longevité

To the memory of my grandfather, David L. Bacon: a man who taught me the value and occasional joy of always knowing to the best accuracy possible where I am and where I'm heading.

ACKNOWLEDGEMENTS

Like most people, I have a day job that demands I focus on just a small subset of the world's problems. In that respect I am luckier than most people for two reasons:

1) My job encompasses the entirety of the human experience as reflected in a microcosm (the International Space Station), and

2) My peers and managers understand that I have needs to fill and contributions to make outside of the job for which they pay me. I wish to thank the people who have made it easy for me to get outside my box routinely, to teach, and to learn at hundreds of events away from my daily work. My perspective of the world would be much narrower, and my life much less joyful, if they ran things by the book. My special thanks to Valin Thorn, James Dunn, Bill Spetch, Jeff Arend, Skip Hatfield, and Mark Geyer for letting me go and for trusting me to make up the time and the work. Thanks also to Max Keller, Robbie Hampton, Linda Ruhl, Harry Johnson, and especially to Eric Schultz for covering so many meetings in my absence.

-Jack

"The rapid progress that modern Science makes causes me on occasion to regret that I was born so soon."

-Ben Franklin to Joseph Priestly

FOREWORD

If you could travel back in time seventy thousand years, the activities of human beings would be scarcely noticeable. Evidence of a genetic bottleneck in our ancestry suggests the human population may have been reduced to a few thousand individuals worldwide, scratching out a desperate existence against the forces of nature. Today, we have transformed the planet. In almost every field of endeavor, with the possible exception of religion, the history of the past few centuries has been one of breakneck and accelerating progress. Advances in science, technology, agriculture, medicine and economics have supported a huge increase in population, a doubling of life expectancy, a surge in the standard of living and some spectacular accomplishments such as the moon landings and the sequencing of the human genome. Many people find it hard to accommodate the rapid changes taking place around them and recoil in what Alvin Toffler famously called "future shock." How many times have you heard someone say, "Where will it all end?"

Futurologists divide into optimists and pessimists. The optimists embrace accelerating change with alacrity; no future shock for them. The future can't come fast enough, with its promise of dazzling new opportunities and products, from robotics to virtual reality, from space colonization to immortality. Ray Kurzweil has coined the term "the singularity" for the time, which he predicts lies but a few decades in the future, when the pace of change accelerates almost without limit, reaching a point where human society is completely transformed into something totally new and at present hardly imaginable. By contrast, the pessimists think that mankind will destroy itself or wreck the planet before long. The President of Britain's Royal Society, Lord Martin Rees, fears that this may be our final century.

Lurking beneath this schism is a split between those who see accelerating change as inevitable, akin to a law of nature, and those who believe that the past five hundred years have been an exceptional episode in human history. The disagreement is well exemplified by what is known as Moore's Law, enunciated decades ago by the co-founder of Intel, Gordon Moore. It states that computer processing power will double roughly every eighteen months. Mathematicians call this rate of growth "exponential," and it has the characteristic feature that in a very short time either the growth rate must stall, or something truly dramatic occurs. Exponential, or even super-exponential, growth characterizes many of the advances over the past centuries and decades, and not just information processing. It is easy to do the math and figure out what exponential growth rate implies for ten, twenty, thirty years hence. And there is no doubt that if information processing power continues its dizzy spiral, some mind-boggling possibilities lie just around the corner.

But suppose Moore's Law fails? What if the exponential growth rate of the recent past is an anomaly? Some skeptics foretell the end of the great leap downward to ever smaller and faster processors. Unfortunately, a fundamental limit is imposed by physics when the components approach atomic dimensions, and quantum uncertainty is encountered. But the promise of the so-called quantum computer, in which quantum mechanical effects are exploited as a virtue rather than evaded as a sin, could represent as big an advance in processing speed and power as the original electronic computer represented over the abacus. At this time, however, there is considerable controversy over whether quantum computation is an achievable goal.

Jack Bacon is definitely on the side of the optimists. His work as a systems engineer has taught him the importance of creative synergy. When human command and control merges favorably with customized hardware and software, the result is much more effective than individual components operating separately. Bacon foresees the virtuous and mutually reinforcing confluence of several rapidly improving factors: faster information processing, a better educated and more inclusive workforce, longer life expectancy and enhanced mental performance techniques, the growth of the web, and the burgeoning development of intelligent systems. The resulting "combinatoric explosion" will, he believes, open up extraordinary possibilities not only for technological marvels beyond our wildest dreams, but for understanding the world about us, including the origin of life, the structure of the universe, and the principles governing complex systems.

Viewed on evolutionary or geological timescales, the human drama is virtually instantaneous. Within the twinkling of a cosmic eye, a single species has figured out the rules on which the universe runs, and developed technology so sophisticated that in many realms it is challenging nature for supremacy. Who can say what our destiny will be? Will the coming centuries see mankind spread into the solar system and beyond, or will our fragile biological intelligence give way to a more robust form of machine intelligence? Are we at the threshold of a momentous elevation in our collective fortune, or are we *homo sapiens* engineering our own demise? Whatever your conclusions may be, this book will leave you exhilarated, fascinated – and perhaps a little scared. The future is closer than you think.

Paul Davies
Sydney, February 2006

Dr. Paul Davies is a cosmologist, physicist, Professor and Director of BEYOND: Center for Fundamental Concepts in Science at Arizona State University, an internationally known media spokesperson for scientific efforts, an author of more than twenty-seven books on science and its impact on society, and is a winner of the prestigious Templeton Prize. See:
http://aca.mq.edu.au/PaulDavies/pdavies.html

OUR PARALLEL UNIVERSE

Look around you. Every modern item in your environment either has, or had on its packaging, information that helped people to track it along the factory floor, into the shipping crate, into the seller's warehouse, onto the shelves, and through the cash register. Barcodes and serial numbers are now part of almost every commercial product. Information about the physical object is on shampoo, catsup, dog food, computers, automobiles, this book…just about everything. Even our fruits and vegetables now have numeric codes plastered (and soon, tattooed) on them. This information wasn't always there, even half a century ago. Until recently, humans laboriously transcribed a representation of every product and every possession into their minds and into their paper records. We found abstract ways to represent the physical world in a parallel world of information that only humans occupied (or even cared about).

Wander by a telephone pole or a fire hydrant sometime and look for the little metal plate or stamping that identifies that pole or hydrant uniquely in the world. You are uniquely identified too. Your social security number and other codes correlate you and your activities with other facets of the world, such as your work hours, pay, and vacation time. As you drive along, your cell phone company can instantly find you (even overseas) and route incoming and outgoing calls through your miniature communicator—just like they did on *Star Trek*.

Location information is not just limited to your phone. The proper authorities know where your car is. Drive through a toll booth, and know that you'll get the bill at the end of the month. In the new parallel universe of information, electronic bits of information record that in the physical world you owe the road service a certain amount for a physical event (your toll booth crossing) that took place at such-and-such a time. The bill for this transaction gets submitted in the form of electronic bits from the toll authority's computer to your bank's computer, and more electronic bits are sent from the bank to the toll authority, thus enriching the latter with electronic representations of your (now diminished) worth in the real world. The whole transaction takes place in a parallel universe of information, without your physical involvement after the initial event. In fact, you can pretty well live your life without physical cash. You need merely identify yourself to the store with your credit card, and money is transferred in the *parallel* world while inventory is depleted there too. Even if you still use cash, notice that each paper bill you use has a serial number on it. We assign information to *every* item.

We've always invested some of our physical efforts into transcribing real-world events into a form we can manipulate mentally. We'll soon see that an amazingly disproportionate share of energy is consumed in the region of the brain where this work is done. Even before the electronic age, we'd always kept inventory records. In fact, inventory is most likely the reason we have writing at all. We've always had a parallel abstraction of the physical world. It used to take more effort to record it than it does today, but it has always been worth our while to spend physical and mental energy to model our world. It's what we makes us unique among the animal species.

Our physical investment in time and labor for transcription to the parallel universe used to be a more substantial portion of our day. Because the abstract representation can so easily get out of synch with the real world (lost or stolen items, lost sales slips, corrupted data, etc.), we had to spend days or weeks every year in every business, just counting stuff and logging it in to our parallel model of the actual store and warehouse stocks. Barcodes and portable barcode readers made a huge dent in labor to keep track of inventory, but you still had to get right next to an object in order to transfer its existence and location into the mental model. These days, Wal-Mart, the U.S. Department of Defense, and others are introducing the consumer to Radio Frequency Identification (RFID) tags that make inventory *automatic*. Data is shared at a distance from the product itself to the store's inventory system. You no longer have to look at an item to count it, or even to find it.

Efforts are underway right now to make the retail cost of such automatic information aids less than a penny each, or a fraction of a percent of the value of the physical item.

Of course, raw data and counting cars and soup cans is one thing. It is quite another thing to be able to *predict* the behavior of physical world objects. Your passing through the toll booth added a bit of knowledge to the folks at the highway department. These people measure how many automobiles are using the roadway, say, on Fridays at eight o'clock PM, as they ponder the optimum time to close lanes for repairs. Instead of just representing static objects, we are creatures that specialize in *predicting* what will happen. Our brains have unique large frontal lobes just so that we can make our sophisticated plans. Recently, though, we've done a lot more than just out-think herds of wild game, or strategize when to order more cream-of-mushroom soup, or to close a highway lane.

In 1589 Galileo Galilei began to model the physical world in mathematical equations: abstractions and approximations that could represent (and more importantly, predict) the physical performance of simple effects in the real universe. Later, Kepler, Newton, Mendel, Maxwell, Thompson (Lord Kelvin), Curie, Schrödinger, Rutherford, Einstein, Feynman, Watson, Crick, Franklin, Hawking, Veneziano, Venter, Smalley, and thousands of others added to the puzzle's solution, incorporating better and better abstract representations of the real universe and of all its inhabitants into the model. That model is our parallel universe of knowledge and understanding.

Close your eyes and look around yourself again from deep inside your mind. It's amazing to see how much you know for a fact, that your ancestors of only a few generations ago (and all that preceded them) could only wonder. In your mind's eye you actually *know* where you are: both in the universe in general, and also relative to many tiny facets of your own body. You're confident that you're on (or over) some point on the globe that you understand pretty well, from its continents to its seas to its atmosphere, etc. Depending upon your astronomy training you can also probably pinpoint where you are located within the Milky Way Galaxy, and you're probably aware of the unequivocal evidence that there are tens of billions of other galaxies out there, and we're all flying away from each other at a very fast but slightly decreasing rate.

You're aware of your physiology, and that salt raises your blood pressure, and that saturated fats are bad for your arteries, and that you are made of many cells that grew from a single cell that formed at conception, and that patterns of four key bases along a chemical strand called deoxyribonucleic acid (DNA) govern the large-scale operation of all cells (and the animals and plants that are made from them). You're aware that there are atoms that make up the known elements, and that electrons, protons, and neutrons make up atoms, and that there are subatomic particles that in turn make up these building blocks

Now "we"—that is, *some* folks among you, me, and our other fellow humans—are closing in on the "Theory of Everything." We are developing new mathematics to deal with the startling vistas that open to us as we explore the world that no animal but *us* has ever seen: complex interactions among the smallest particles, the most intricate and ethereal of forces, and the content and history of the entire cosmos. We now have a complete map of the chemical code that defines us, and we are busy building maps of how our brains are wired. We track how the different regions interact as we think, sense, and emote. We have the model in place, and are busy polishing it into a more perfect representation of the physical world.

What makes this all the more astounding is the exponential rate at which this parallel universe of ours—our mental representation of the real world—has coalesced to an ever-simpler set of rules and then explosively expanded into comprehension of the entirety of the physical world. The human mind now comprehends its physical domain from the smallest particles to the size and workings of the universe itself, and it has done so in an almost infinitesimal time.

THE PARALLEL BANG

Examine the figures on the following pages. The areas of each figure depict the relative time scales of our universe and of our planet, and in the second figure, of human intellectual growth. Several things are of note. In Figure 1 the scale of the area of a page represents the time that our universe has existed. We humans have been around for a period of time of only a million years: less than the space devoted to the letter "i." Moreover, on that scale all spoken language, cave painting, and human tools have existed only in the time represented by the dot on the top of the "i"—about seventy thousand years.

To fully come to grips with the tiny period of our true intellectual growth, we expand that area equal to the i to be the size of a page, as in figure 2. The era of modern science, since about 1800, is once again the size of an "i" on this new page. Considering that the truly explosive growth in science and understanding has occurred in just the last fifth of the time since then, and that this infinitesimal slice of time is *also* the time when all intellectuals on the planet have linked through the modern communications network, it's clear that something big is happening.

Next 2 pages: Figures 1 and 2. The relative timescales of the universe are shown as areas of sequentially smaller boxes.

Age of the universe (13.7 billion years)

Our Earth and sun formed 4.3 Billion years ago

Microbial life formed (3.8 B years ago)

Death of the dinosaurs (65 Million years)

Homo Erectus (ONE Million years)

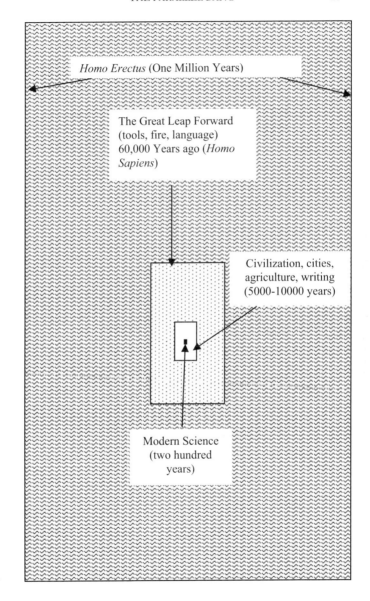

That final period, smaller than Figure 2 can show, is also the period when all previous writings and models have been archived and are instantly accessible through the web. It is also the time when the real-world's temporal information, such as global weather, ocean currents, rainfall, and agricultural resources (not to mention the inventory of all man-made goods and produce) has moved into the parallel realm. You can sense that something truly amazing is afoot in the physical universe (or at least in our little corner of it).

Our intellectual, parallel universe has exploded into galactic scales out of near nothingness in a tiny sliver of time: a birth similar to the big bang that started it all: a parallel bang of human understanding. Moreover, it's not just computing that has propelled us forward. We have realigned our cultural priorities to put an ever-increasing portion of our societal energy into intellectual efforts. A large and growing fraction of our population, both in the sheer numbers of workers and in increased diversity of gender and race, is at work on intellectual issues. In the coming pages, we'll explore further the explosive growth of our fledgling parallel universe, examine its causes, measure its expansion rate, and see what effects this expanding new parallel world may have on our real-world lives.

THE PARALLEL BANG

As a technological historian, I have studied the pace of change of human society over the last thousand years. As is evident from Figures 1 and 2, this is a miniscule portion of natural history, but it covers all of the growth of human thought and philosophy since the Dark Ages. Having been privileged to learn of and to explore twenty-eight generations of my own ancestry, I took it upon myself in 1998 to chronicle the changes that my family had seen from generation to generation since the Battle of Hastings in 1066. Until the time of the Renaissance, it was easy to pick a day in the life of each of the ancestors that was typical of their time and which showed the slight changes in theology, technology, and society from the previous generation.

However, since the time of the Renaissance, the challenge of capturing a single day in the life of any of my ancestors grew with every generation, as society changed many times within each lifetime. Now, even that high rate of change is taking an explosive upward turn.

Until the time of the Renaissance, human societies doubled their technology and knowledge of the world roughly every three hundred years. Occasional cross-pollinating events (such as Marco Polo's return from China in 1295) caused local accelerations of understanding and knowledge, but on average, we muddled along by doubling our capability every ten generations or so (which is pretty astounding, when compared to, say, chimpanzees).

At the beginning of the Bronze age, we achieved several new capabilities. Writing enabled transport of thought across generations and terrain. Sails connected individual travelers and merchants across vast waters every day. Agriculture enabled royal bounty that in turn engendered courts where there was time to think and create, rather than to just survive. Humans all of a sudden went from doubling our technology every two to three thousand years to doubling it every two to three hundred years.

At the beginning of the Renaissance, human society (or the part of it that subsequently dominated world culture) achieved several more astounding enablers. The movable-type printing press allowed mankind to record for all time (and in numerous perfect copies) the accumulated knowledge of the ages and to share that knowledge widely. The caravels that Columbus sailed to the new world in 1492 were typical of the commercial ships of the day. These were just starting to visit every port daily cross-pollinating all the major nations of Europe. These ships brought more than goods: they brought ideas and books and philosophy. This promoted a flowering of human intellect, riding on the waves of commerce. The voyages also brought an abundance of new information about previously unknown lands and species far over the horizon, challenging millennia of established (but erroneous) thought. Lastly, within a hundred years of the start of the Renaissance, Sir Francis Bacon established the scientific method, the process by which humans could recognize what they did not yet know, and find a way to efficiently resolve their ignorance with established fact.

With all of these innovations, human society moved from a technological and social doubling time of three hundred years to a doubling time of only thirty years, or roughly once per generation. Humans were learning as much in a decade as they previously had in a century or more. The western world began an astounding rise in intellectual capital and comprehension that propelled a deep (and ever deepening) separation from the world's tribal cultures.

The accumulated knowledge of weapons, transportation, and growing wealth of well-organized nations led to standing armies and trans-continental conquest. The advanced knowledge and tools also led to a total mismatch of capability in every encounter thereafter between the technological and tribal cultures, as the world began to re-unify. Further, the long exposure of western civilization to numerous domesticated animals led to a plethora of diseases to which westerners were immune, but to which the other cultures of the world had no resistance.[1] The

---

[1] For a compelling explanation of the dominance of the western European culture in a world populated by an

advantage of even a few generations of technological doubling was enough to overwhelm any less advanced culture. As Arthur C. Clarke has observed, any technology significantly further advanced beyond your own is indistinguishable from magic. Magic is hard to defeat.

The average citizen of today now lives several hundred times better than did his or her ancestors of only five hundred years ago, while our few remaining tribal brethren are still clinging to the unchanged ways of the ancient ones. Virtually every part of our society in medicine, transportation, agriculture, astronomy, chemistry, and every engineering and academic pursuit has—in the words of Isaac Newton—"built upon the shoulders of giants." For half a millennium (or one percent of our *Homo Sapien* existence), humans have been exponentially adding to the body of our knowledge and to our understanding of our place in the world and in the universe.

I propose that the intellectual sparks of the last five hundred years have been only the burning of the fuse on a very large bomb —an explosion of intellect that is just detonating in this generation. For millennia before, we were somewhat stable in our species. In the last five hundred years, we've seen a wave of intellectual energy and discovery trigger a spurt of human understanding, as well as global unification. I believe that we've turned a

---

otherwise genetically identical species of humankind, see Jared Diamond's *Guns, Germs, and Steel*.

new corner—that this burning fuse has kindled a new thirty-fold explosive growth rate of human intellect. When compared to the already sudden growth of this knowledge-based species in the natural history of the world, our nearly instantaneous new growth is something of truly remarkable significance to this planet, and perhaps to our region of the galaxy.

We humans are the only type of animal to adapt the environment to ourselves: all other animals must wait for random genetic mutation to allow some of their descendants to better fit their ever-changing environment. We humans are the only species to live from pole to pole, under the sea, on the surface of the oceans, under the earth, on the land (at nearly all elevations), in the sky, and now even in that most hostile of environments: the vacuum of space. We will inhabit other planets and moons of our solar system within a few decades. Interestingly, humans are among the frailest of all mammals. We are neither as strong, nor as fast, nor as resistant to disease, nor as impervious to the extremes of temperature, of oxygen, nor of any other environmental threat as our mammalian relatives. The reason that we thrive is our ability to understand our environment and to react to it on an intellectual and technological level.

In effect, we've chosen (or have been chosen by our genes or by God) to evolve in our parallel universe of understanding. All other species react genetically (and slowly!) to the forces of the physical world. We humans take that same physical world and represent it as information. Our data, analytical models, philosophies, physical principles, etc, are a sketchy representation of the real world: a parallel world maybe, but a type of world that we alone occupy, and a world where changes are rapid—sometimes explosive. We can re-pattern our minds *billions* of times faster than we can re-pattern our genes.

In our parallel representation of the world we take intellectual journeys to the depths of subatomic particles and understand their invisible, mysterious forces. We capture global weather patterns, hurricanes, tsunamis and tides in our feeble brains. We come to grips with the minutiae of cellular biology, and the chemical and physical processes by which we reproduce, grow, live, and die. We've decoded our genome—the ultimate root of information about our existence. We explore planets, stars, galaxies, and the universe itself in our imaginations, supported by a growing body of data from our technological tools. We even create imaginary dimensions in mathematics and physics (where, for instance, the square root of minus one has real meaning and value to our understanding of the world we sense).

With every piece of the puzzle, our parallel universe of models, thoughts, and understanding becomes more complete. From the cosmic to the atomic scale, we have the entirety of the physical universe re-forming in our heads. Everything about our human society is now modeled, analytically stored, and digitally abstracted in a world of mathematics and intellectual representation in every way parallel to the physical world. We create virtual reality and simulated worlds where every facet of our lives is modeled, recorded, stored, and shared. Of critical importance, with those models of the universe we *decide* our next evolutionary steps. From the Domesday Book to today's modern census to barcodes on our shampoo, we measure every part of our real existence in our parallel universe of information and knowledge.

I believe that we can create an estimate of the future rate of growth of human understanding, or at least conclusively demonstrate that such a rate will be significantly faster than it has been in previous eras. We will do this by examining the historical record for the causes of human intellectual growth. We will then look at the factors that are at work today, and that are adding to the present acceleration of understanding. This approach will give us a basis for estimating the expected effects. Finally, we'll link brains with many others via the web, to reach a consensus and to continue the dialogue.

The indicators of our new growth are all around us, and we can measure them, giving us some idea of the new pace of human intellectual development: the explosive growth in human understanding that I call *the parallel bang*.

GROWTH SPURTS

The First Wave: 65,000,000 Years ago

As Carl Sagan observed in *The Dragons of Eden*, mammals were the first creatures capable of storing more information in their brains than could be stored in their genetic code. This anatomical feat occurred essentially coincident with the demise of the dinosaurs, sixty-five million years ago. (See page 20 for a sense of how recent that was!) In other words, when some little mouse-like creature developed a brain that could store seven hundred fifty million bytes of information, the race towards a truly intelligent being—in whom the whole physical word could be re-created in information—was on.[2]

---

[2] The abundance of intelligent life in the universe is a subject of hot debate, but fueled by little data at the moment. It's tough enough to establish the fine details of intellectual growth in Earth-based species (where fossil records are available) without worrying about life elsewhere, where so far, such records are not available. However, a statistical case is to be made that intelligent life is highly abundant in the universe, using the famous Drake Equation. Another case can be made that some key features, including star size, neighboring stars, existence of a planetary molten core, narrow range of allowable planetary distance from the star, and many other features are low-probability preconditions to the rise of anything above single cellular species. For a critical, analytical look at the potential abundance of intelligent life, see Ward, P.D., and Brownlee, D: *Rare Earth* (New York Copernicus Books, 2000)

Sixty-three million years later (or two million years ago), the evolutionary path that emphasized intellectual prowess had arrived at *Homo Erectus*: a walking man with a brain quite similar in size to a modern human's, but lacking certain key features, such as fully-developed frontal lobes. However, *Erectus* had all the hallmarks of being a good pattern-matching machine, with vast amounts of memory—memory that had to be patterned anew in each new generation.

The Second Wave: 60,000 Years Ago

The Upper Paleolithic Explosion occurred one hundred thousand years ago. That's when we got cave painting (a primitive form of history book), body ornamentation (use of technology to enhance personal standing), enhanced tools, ritual burial, and trade over long distances. It appears that spoken language was now part of our species, and (evidenced by ritual burials) it appears that we had a collective model of the universe with future implications—a collective model shared and practiced by large groups. In short, we had new tools for recording information, creating new ideas, considering the long-term consequences of our actions, and, via trade, of sharing each idea over wide areas. All of these were signs of markedly better pattern building than existed in other primates.

Slight genetic variations led to competing branches of *Homo Sapiens*: Neanderthals and Cro Magnon Man. Cro Magnon ultimately won out, mostly because the organization of Cro Magnon's brain was slightly better adapted to a broader range of problem solving.

As Cro Magnon emerged victorious, the Great Leap Forward of 60,000 years ago indicated a final knitting of the human brain into the structure we know today. Since then there has been little genetic or archeological change in human anatomy, and little physical change in our brains. (This is partly limited by the ability of women's hip bones to allow both routine erect walking and the periodic birthing of a child with the largest possible cranium. At three times the normal brain-to-body weight of our nearest primate relatives, we've apparently hit an optimum for head size at birth and sustainable biophysical stature throughout life. Even so, a human infant is born with relatively soft cartilage in the skull that must expand quickly in just the first two yeas of life, ossify, and knit together into the bony skull we find in adults.)

Today's human brain can store approximately fifty trillion bytes of information, or sixty-seven thousand times what our DNA can store. (This is also approximately fifty times the equivalent data in the U.S. Library of Congress). We (and all mammals) nurture our young for long periods, transferring models of the universe into those comparatively blank minds. Because humans have the greatest investment of all mammals in the brain's role in the species' evolution, we spend the longest relative portion of our natural lifespan caring for our young. During that time, we concentrate on transferring information.

We enabled our ideas to outstrip our genes sixty-five million years ago, and settled on our current balance of brain size sixty thousand years ago (with the arrival of *Homo Sapiens*, literally, *Wise Humans*). We have since that time done our principal evolution as a species simply by building better and better patterns within that "standard brain" as each generation went by. Not better brains—just better patterns.

The Third Wave: 5000 Years Ago

Five thousand years ago we had migrated to almost every arable part of the world (except to New Zealand, which was settled less than one thousand years ago, Chatham Island six hundred years ago, and Pitcairn Island two hundred twenty five years ago). In certain parts of Eurasia, humans ended the Stone Age and entered the Copper Age (followed by the Bronze and Iron Ages). For reasons outlined in Jared Diamond's *Guns, Germs, and Steel*, these inventions (i.e. ideas) propagated east to west throughout Eurasia, but not across the now-lost land bridges to the more dispersed members of our species, and not southward to Africa across deserts that could not sustain an agrarian lifestyle.

Key ideas like metallurgy were possible in Eurasia mostly because of the time and energy that humans had made available to themselves for intellectual pursuits, available only because we now had the first writing, formal bureaucratic social structure, agriculture, animal domestication, mathematics, astronomy, plumbing, etc. Metal smelting was just a small (but highly-enabling) part of the intellectual achievement of the era.

We modeled the physical world, represented it on papyrus and clay tablets, transported, shared, and stored these representations, and began real engineering of sophisticated tools and buildings, not to mention effective social structures. We exercised dominion over a few plant and animal species.

Our population began to grow, especially in areas where technology such as agriculture, metalworking, and ship-based trade supported healthier, more abundant diet and a less risky livelihood than hunting and gathering. The abundance of agriculture allowed society to support a few of its members in duties not related to the production of food. For the first time, we assigned a dedicated portion of our number to the social role of organizing, planning, divining the will of our gods, documenting our productivity…in short, we set aside social resources to the tasks of pursuing thoughts about our existence, and how to improve it. These individuals were tightly clustered around the leaders of society. Information *was* power.

The Fourth Wave: 500 years ago

Nearly five thousand years after the third wave, The Renaissance arrived. We quite suddenly acquired books, intercontinental ships, and the scientific method—all within just over a century. These boiled down to better recording of ideas, more prolific sharing of these ideas, and better means to find new models or to refine old ones. Of special note, these inventions allowed a far larger portion of the population to engage in intellectual pursuits. Over the following centuries the creative engines of society moved swiftly out of the courts and into the population at large. Inventors and tradesmen amassed great fortunes and new power based upon their ideas. After considerable upheaval, the common citizen was allowed to read the Bible, and it and other documents in archaic languages were translated and dispersed to the masses. For the first time, women were allowed to read. That doubled the available intellectual capital in a single move.

European explorers rediscovered our long-dispersed brethren in vastly different cultures, and came to comprehend the size of our Earth, our place in it, and its place in the heavens. This in turn shattered millennia-old beliefs and spawned new philosophies. The Christian Church was rent asunder numerous times as Catholics split from Eastern Orthodox and Russian Orthodox, Lutherans split from Catholics, Methodists and Congregationalists from Lutherans, etc. New scientific answers challenged millennia of thought patterns, and with every new philosophic variation, debate and the pursuit of more new answers (to resolve the new debates) intensified. Ideas evolved with stunning rapidity in every realm of human effort and thought.

Wars began to shift from struggles for land to struggles for ideas and ideology, such as the American and French Revolutions. Human existence started to focus increasingly on our thoughts and less on our resources, as our ability to engineer the environment to better sustain ourselves took more and more physical hurdles of subsistence out of the way of intellectual growth.

The Fifth Wave: 5 Years Ago

Most fans of modern technology are keenly aware of a phenomenon known as Moore's Law: the observation (first made by Gordon Moore in 1965) that computing technology has doubled its performance per unit of silicon area every eighteen months: in other words, every five hundred forty days. It's been true for nearly fifty years.

We'll get into why Moore's Law is true in another chapter, but for the moment, consider this: The computer (in any form like the version we currently know) did not emerge until fifty years ago, and really only found its way into individual consumer hands twenty years ago. The average personal computer now processes information faster than a guppy but not as fast as a mouse. It will outstrip the mouse in fifteen years. By

Moore's Law, your home PC will be able to out-process your own brain (in sheer numbers of synaptic firings per second) within your lifetime (by 2040 on the current rate: some say much faster than that). You can now carry more information in your Blackberry®, Palm Pilot® or iPod® than you do in your genome, and will soon carry more in those devices (or in your glasses) than you do in your brain.

Let's recap: under the first wave it took 3.8 *billion* years develop a pathway other than chemical genes for information growth in a species (brain to brain rather than gene-to-gene). It took sixty-five *million* years (less than two percent the wait time for the first wave) for that brain to develop the second wave's alternate channels (painting, technology, trade) that enhanced the information transfer many times above that of other mammals. It took *one hundred thousand* years (one seventh of one percent of the previous wait time) to get to the third wave of the Copper Age, agriculture, writing, and city dwelling. It took five hundredths of the previous amount of waiting time (*five thousand* years) to create the tools of the fourth wave (the Renaissance).

Now we've had an unbelievably long relative wait time: *five hundred* years (or a full tenth of the previous wait), for the fifth wave to begin, but it most certainly has begun. In the last decade, we've launched into a whole new era of human intellect and understanding. Consider the tools at hand:

1) We have the computer to sift and to store information.
2) We have the World Wide Web to share it.
3) We have nearly free voice and video links to every intellectual on the planet.
4) We communicate more information every year than our entire species has ever spoken aloud in all of our two million years on this planet, and that rate is growing geometrically.
5) All archived text in the world's major libraries (i.e., everything worth keeping that has ever been printed) will be electronically searchable by the end of 2010.
6) We are (or soon will be) able to remotely locate and to effortlessly count every material possession and every relevant exchange of goods and services in our parallel world's databases.
7) The languages that divide us are unifying, not only because thousands of tribal dialects are going extinct each decade, but also because the major languages all include all modern words in nearly the same form: *Radio, Antenna, Computer,* and *Tokomak* are all universal, for instance. In addition, automatic translation is here, and getting better each year.
8) Our models of the world are more self-consistent and universal than ever before, spanning the entirety of the known scales of the universe. Our mental artistic

representation of the real world's laws and dimensions has gone rapidly from Abstract to Cubist to Impressionist to Realist to the image quality of HDTV. There's not much anymore that isn't accurately rendered on our intellectual canvas.

9) We have instant search tools in our electronic infrastructure to find the missing pieces, to find resources, and to organize the search for new knowledge.

10) Eleven times as many college graduates are contributing their creative efforts to society as were contributing only forty years ago. Annual patent filings *per person* have more than doubled in the last ten years, and annual earned PhD degrees have more than doubled in twenty years.

11) There are efforts underway to provide the modern information tools to *every person on the planet*.

And that's not all. There are other factors at work that are helping to accelerate our accumulation of knowledge in the parallel mental universe—factors that we'll explore soon. For now, it should be clear that we're on to something big. Don't blink. The world is about to change.

THE BASIC EQUIPMENT

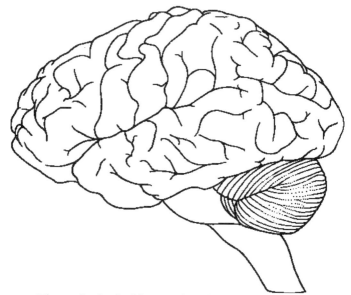

Figure 3. Left side exterior view of a human brain. The function and chemical agents involved in each visible region in this and the next diagram are understood to a large degree, with much work remaining.

I've asserted that our parallel universe exists in the human brain, so let's spend a little mental effort exploring this organ, to see how it works, and how it can possibly hold so much information.

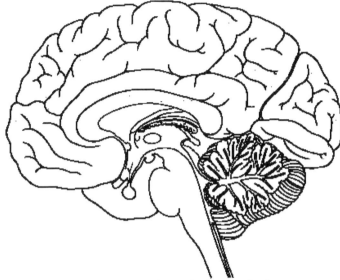

Figure 4. Cut-away view of a human brain.

Proportionate to our body mass, the human brain is three times larger than it is for any other primate, and proportionately larger in primates than it is for any other species on the planet. Still, it's pretty small: about the size of a grapefruit. Put your fingers from temple to temple, and then draw your hand back and examine the size of your outer skull. Shrink it in each dimension by about the width of a finger (accounting for the five layers of protective matter between your brain and the outside world!) and you have an idea of how small this amazing organ really is.

It weighs about 1.3 kilograms (about 2% of an adult's body mass), consists of seventy-five percent water, and has the consistency of soft butter. It is made of neurons and supporting fatty tissue, and contains about one trillion neurons (and ten trillion supporting glial cells), each neuron making approximately one thousand connections, for about five hundred trillion connections or synapses, which can each be thought of as a "bit" of information. Note that these "bits" are not pre-ordained. Neurons are genetically destined to occur, but synapses must be patterned in life.

Like the human immune system, the brain is only endowed with the ability to create useful patterns that fit the environment. In the case of the immune system, the body makes specialized protein patterns on white blood cells to chemically match to infections to which each body is exposed over its lifetime. (This is what immunization is all about: we "teach" our bodies to make the most useful patterns of protein receptors on white cells.) In the case of the brain, the body makes connections between neurons in response to stimuli from the outside world. Most of those stimuli come from our parents, teachers, and peers, and this process is not random. We teach our children (or they teach themselves by experience) the patterns that are of greatest use in real-world events. Making each new pattern requires expenditure of energy. Once the pattern is built, it takes very little energy to re-match the pattern to a stimulus. (For example, a green light means "Go" and a red light means "Stop." No need to solve new problems.)

Thus, your brain holds about fifty thousand CDs worth of data that must be reprogrammed from scratch in each individual. Remember that a good encyclopedia with video and audio clips fits comfortably on one CD. If you doubt that you have this much stored up there, ask yourself how many scenes of motion pictures or pop song snippets you can recollect. It takes only a few seconds of a video scrolling by on television for you to recognize that you've seen it before. That brief pattern (and all others, from all movies you've ever seen) is stored up in your brain somewhere.

Of special evolutionary note, the brain, at 2% of our body mass, consumes ten times its share of energy: 20% of the body's oxygen is consumed there. Of that huge cranial oxygen supply, 94% is delivered to the thin gray matter of the outer cortex (the 40% of the brain mass that is unique to primates, contained in a highly-folded sheet only a few millimeters thick). The remaining 6% of the oxygen is delivered to the 60% of neurons that are white matter, deeper inside, responsible for basic animate functions, and common to the less cognitive species on the planet. In other words, the uniquely human portion—the gray matter—is consuming *twenty times* its share of the body's energy. Clearly, our bodies have devoted a lot of resources to this evolutionary niche: most of the work is going on in the unique region of our human brain that creates abstract ideas and predictive models, and then expresses them. The non-primate features of the brain are consuming merely the average body share, just like a spleen.

Our brain is divided into layers, with hemispheres in the outer cortex layer (about three millimeters thick, but folded much thicker) with lobes and folds within the lobes. Each physically obvious partition or fold has specialized local functions, and each is able to cooperate with other such regions in more complex thought patterns.

The brain is not one entity, but dozens of essentially independent processors both cooperating and vying for resources. Twelve cranial nerves connect the sense organs (eyes, ears, taste, touch, smell, and major touch and motor centers) into the brain. These cranial nerves are located roughly under the central brain, towards the roof of your mouth.

The *corpus callosum* ("tough body") is a stiff stringy mass between the two halves of the outer cortex. It acts as the superhighway between the two lobes of the outer brain for coordinating abstract thought. (The physical real-world interactions through the senses are carried at much lower bandwidth in the cranial nerves). The *corpus callosum* is larger in women than in men.

Neurons that transmit over longer distances tend to be sheathed in myelin, provided around the long nerve axon by the structurally supporting glial cells. Myelin acts like an electrical circuit accelerator, greatly shortening the time it takes to transmit a signal along the nerve cell. Neurons in the gray-matter cortex do not have as much myelin, and generally are more interconnected with each other than those in white matter.

The two hemispheres of the brain have different specialties in abstract thought, and each controls the opposite side of our symmetric bodies. The left side controls the senses and motion on the right side of the body, and the right side of the brain controls the muscles and senses on the left part of the body. In addition to these basic body control functions, the hemispheres have particular strengths in certain types of thought. The right hemisphere is associated with holistic, integrated "big picture" thought, abstraction, visualization, music, geometry and shapes. The left hemisphere is endowed with higher strengths in detailed analysis, step-by-step thinking, language and vocabulary. Two dime-sized areas of the brain (Broca's Area and Wernicke's Area, associated respectively with speech motor skills and with recognizing the shapes of words and associated sounds) are only active on the left side of the brain, although similar topographic features exist on the right.[3] A "left brained" person might give directions with

---

[3] The left-right division of function in a nearly symmetric entity should not be a real surprise: even in our symmetric bodies, we all have the liver towards the right, stomach, heart, and spleen towards the left.

street names, while a "right brained" person would describe landmarks. Every person's thought patterns are a mix of skills, but generally, one side of the brain out-influences the other. It varies from person to person how unbalanced the dominance is.

Statistically, the hemisphere of the brain that dominates your thought process correlates with your handedness. If you prefer to use your right hand to write, it is probable, but not certain, that you will exhibit thought patterns associated with the left hemisphere of the brain, which generally is exercised more in daily physical motor tasks. Only 30% of people who are right-handed are "right brained," and an even smaller fraction—10%—of left-handed people are "left brained." People who are left-handed seem to balance the functions more evenly between the hemispheres than right-handed people, but do *not* reverse the general left-right hemispheric division of specialized intellectual skills. With great effort, some naturally right-brained functions can be acquired by the left brain and vice-versa, as noted after strokes and other brain injuries. When the *corpus callosum* is cut,[4] disarming effects of essentially two separate minds within one skull are immediately and permanently observed.

---

[4] Sometimes the *corpus callosum* is intentionally cut in patients with uncontrollable epilepsy, which is like an uncontrolled and life-threatening electrical feedback loop across the *corpus callosum*.

Generally, each specialized region or fold in the brain is likely to cooperate with, and be logically related to, the functions of the physically adjacent regions. For instance, the cerebellum at the lower rear of the brain, which coordinates motor skills for posture and balance, is adjacent to the pons in the central rear, where the sensory inputs for orientation (orthostatic sensors, mostly in the ears) are integrated. The cerebellum is also adjacent to the occipital (rear) lobes of the cortex. The occipital lobes specialize in processing the visual fields of the eyes, and are thus also involved in orientation. Adjacent to the occipital lobe, the inferior parietal (back top) lobe is (amongst other things) where we imagine geometries in three dimensions, visualize rooms when we read about them, etc. The right inferior parietal lobe does most of this work. Thus, a logical arrangement of functions in adjacent lobes of the brain is apparent. (Analysis of Albert Einstein's brain revealed that he had somewhat larger than usual inferior parietal lobes, which probably contributed to his uncanny ability to perform *gedanken* ("mind only") experiments. He also had a larger than usual number of structurally-supporting glial cells, which support and feed the neuron structure, and there seemed to be more dendrites (synapses) than average on most of his neurons. He made more connections per neuron than the rest of us, and that probably led to a richer imagination.)

To put an even finer point on brain mapping, neuroanatomists point out that each fold has an exposed top region (a gyrus) and a crevice (sulcus) where it adjoins the neighboring fold. As you might expect, there are slightly different functions in each sulcus and gyrus, adding further insight into the brain's topography.

Neural scientists continue to map the human brain, aided in recent times by techniques such as positron emission tomography scans (PET scans) and functional Magnetic Resonance Imaging (fMRI) which respectively record the oxygen consumption rate and the electrical firing patterns in the living brain while the test subject performs various mental functions. We'll talk more about these tools later. To date, anatomists and modern neuroscientists have catalogued and understood the basic functions of over two dozen unique processing areas of the brain, but I suspect that few would be surprised to find this catalogue growing in the coming years.

The brain is bathed in clear cerebro-spinal or limbic fluid, traveling through a series of four ventricles or passageways from a central aqueduct, replaced at the rate of about four times per day. The limbic fluid provides a buffer between the brain cells and the body's circulatory system (blood cells are too large to migrate through the intricate neural web), and maintains the suspension of over one hundred unique chemical agents (neurotransmitters and neuroinhibitors). Several glands are *inside* the brain, creating hormones that transfer to the bloodstream and regulate distant reaches of the body.

We are slowly starting to understand what many of the brain's chemicals are and what they do: dopamine shortens attention span, serotonin lengthens it. Oxytocin, (like endorphins, it helps you to feel good) adrenaline, cortisol, acetylcholine, and noradrenalin all affect us. One of the more familiar commercial sleep aids— melatonin—is one of the brain's chemicals, and is naturally produced in the *epithalamus* ("above the thalamus") at the start of the sleep cycle. It is now synthetically produced, sold over the counter, and is my favorite sleep aid when traveling across many time zones: I arrive fully rested and without the grogginess associated with other non-cerebral chemicals. Psychopharmacology is very big business. Such chemistry triggers huge swings in society with every nuance of understanding about the brain's chemical processes. When researchers discovered that the drug Valium interfered with the neurotransmitter gamma-amino butyric acid (GABA: associated with stress), it led to a lot of sales: eight thousand *tons* of Valium went into the world's brains in 1977. L. Dopa, which metabolizes to dopamine, has become a major treatment to counteract the effects of Parkinson's disease, but there is a lot left to learn. The point of this book is that such knowledge, and the benefits from it, are coming orders of magnitude faster than they ever did in the past.

The outer cortex (cerebrum) is where we find conscious thought. The frontal lobes, especially, are where plans are made, the real world is modeled, and problem solving occurs. Although we'd like to believe that we're in control of our own thoughts, there is no central processor within the brain directing the different specialized regions. Each region, it appears, is processing independently, occasionally promoting some out-of-the-ordinary sensation to the cortex for more focused review, in case a plan needs to be adjusted. Some of these "notice me!" features are hardwired, such as sharp pain, hunger, the smell of smoke, sudden movements, loud noises, and the sight of long thin objects in the grass or in trees. As in any bureaucracy, a lot of routine information passes from brain department to brain department without much executive intervention. The chemical soup, if slightly imbalanced, can greatly affect the efficiency and balance of this unconscious information exchange. A shock or traumatic stress situation, for instance, can leave elevated levels of cortisol in the brain for days or weeks, affecting the entire mental balance long after the initial event.

The mind, it appears, is an emergent property of the massively complex interaction of all of the brain's components and regulatory chemicals. As in a thriving ant colony of simple ants, intricate patterns of the colony itself evolve from the much simpler beings within it. One ant's brain is tiny, but from tens of thousands of very simple independent, simple brains, complex patterns of tunnels, foraging, food storage, garbage collection and disposal, and other signs of organized behavior emerge. No one executive ant plans this out…it just emerges as a complex, beautiful pattern. We'll come back to emergence and complexity in the closing chapter. For the moment, it is good to understand that the brain is beautiful, modular, and yet massively complex, understood more every year, and yet posing boundless challenges to those who wish to comprehend its full capabilities and workings.

For further exploration of the physical map of the brain, explore the Harvard Medical School interactive website of the Whole Brain Atlas:

www.med.harvard.edu/AANLIB/home.html

A wealth of additional brain facts can be found at:
  http://faculty.washington.edu/chudler/facts.html,
  http://members.aol.com/Bio50/LecNotes/lecnot25.html
  http://apu.sfn.org/content/Publications/BrainFacts/brainfacts.pdf

Take some time to really explore this amazing organ, because what is about to happen to it is mind blowing. And while you ponder the brain, recognize that yours is the *first* of the last two thousand generations of brain that knows even this much about itself. Only one hundred years ago, we were practicing phrenology, and three thousand years ago, Egyptians didn't preserve the brain in a mummy's canopic jars, because they couldn't imagine what it was for.

THE NUMBER & DIVERSITY OF BRAINS

Clearly the brain is a very complex organ, and within that complex structure, each part is subject to small genetic variations, just like any other part of the body. It would be absurd to believe otherwise, if every other physical trait of a human body obviously varies from person to person, whether it be stature, facial features, hairiness, pigmentation, or metabolism. Clearly, the brain's scores of physical attributes are also subject to genetic variation.

As we've seen already, it is possible to detect variations from the norm even in different "normal" brains, such as Einstein's more intricate inferior parietal lobes, or an average woman's larger *corpus callosum* and higher levels of serotonin than an average man's. Brain disorders like Alzheimer's and Autism appear to have genetic predispositions.

Other than gross anatomical brain differences in men and women, there have been few studies of genetic differences among individuals to correlate with, say, race, body size, or blood type, both because of the ethical dilemmas such work engenders, and because the physical differences are actually quite small and hard to pick out of the noise. The concept of intelligence is so multi-dimensional that it serves little purpose to try to identify strengths (or worse, weaknesses) of any particular group. Compared with the more dominant effects of nurture and environment, the genetic effects on brain development are actually quite small. Remember that humans differ from individual to individual by 0.1% of our genetic code. The biggest single genetic difference is the presence of the Y (male) or X (female) pairing with the other X chromosome at conception, at chromosome pair number 23: a full 2.2% of the genetic makeup, or twenty-two times the possible 0.1% contributions of the rest of the genetic structure. So, for the purposes of this book, I'll encourage you to accept that everyone has strengths and weaknesses, that the biggest general differences lie between men and women, between left-brained and right-brained people, and that there are probably scores of other cognitive variations that correlate with other genetic traits. These other genetic differences are small, however, and add to the richness of a society by allowing alternate points of view.

I know I'm treading on dangerous ground here, for after all, Harvard President Laurence Summers got into some considerable hot water for carrying conclusions about brain anatomical diversity too far. In a 2005 speech, Dr. Summers argued that the differences in male and female brains are *why* there is social inequality in academia. Right data, but wrong conclusion! The reason is prejudice, which is another way of saying that the ideas resident in the current socially dominant members of our species are trying to perpetuate themselves. It is a product of memes, not genes that modern academia is maladjusted to women and minorities. We'll dive into that topic later, and I'll make the case that diversity in brains is good for our species. In the meantime, being cognizant that one can easily step into unpleasant philosophies, let's look at what we know about diversity in the physical hardware of the brain.

We've all heard of "Type A" and "Type B" personalities, left-brained to and right-brained people, "naturally gifted" musicians and artists and engineers, and the idea that men are from Mars and women are from Venus. I am certain that I am not cut out for certain jobs in which others thrive. I know for a fact that I simply don't have the temperament to do repetitive work, nor the patience nor skill to be a master craftsman in any particular art. (Too much dopamine, I guess.) I stink at organization and at spelling. My patience is something on which I need to work. I'm not a perfectionist. My particular skill set includes the traits that I'm a generalist, and that I see the really big picture quite accurately, usually in three dimensions, with multiple, fuzzy interactions of many parameters and influences at once. I get very close to the right answer, very quickly, and then count on the many gifted specialists who work the details to clean up my inevitable mistakes as we polish out a perfect answer.

I'm definitely a right-brained kind of person (although I'm right handed), with a hefty mix of left-brained tendencies, especially mathematical analysis, linguistic capability, and moderately good (but not excellent) logical sequencing of problems. I make a comfortable living because not many people think the way I do. My brain has found an ecological niche in an international space program, where there are many big, integrated problems that need an approximate answer in a hurry. Even though I have a valuable talent in this one important area, my management has to organize my assignments to avoid my known weaknesses and dislikes, assigning such tasks to folks who actually seem to like that work. Diversity is good. We'd be sunk for all kinds of reasons if all brains were like mine, but I like to think that we are better off because there are a few brains similar to mine in the mix. We tend to gaggle together around the same work.

It is interesting to note that in my area at NASA—in a job that requires lots of geometric reasoning and "big picture," non-sequential tasks—we have a disproportionate fraction of left-handed people: several times the statistical expectation of five to ten percent. This is to be expected because of the nature of the "right brained" thinking skills required in our task. Several other big-picture geometric thinkers, listed below, reinforce the hypothesis that left-handedness is indicative of certain mental predispositions. (I can applaud these lefties without bias, because I'm right handed.) Remember that left-handedness occurs in less than 10% of the population and that left-handedness is usually a more balanced weaving of the brain's many functions, rather than a preponderance of left-sided functions. The partial list is as follows:

>   -Four of the last six U.S. presidents.
>   (Ford, Reagan, Bush Sr., Clinton)
>   -Napoleon, Benjamin Franklin, Caesar,
>   - Leonardo da Vinci, Picasso, Raphael, Rembrandt, Escher,
>   -Mark Twain, Leo Tolstoy, H.G. Wells, Lewis Carroll
>   -Albert Einstein
>   -Bill Gates
>   …and Ken Jennings
>   (all time JEOPARDY! Champion)

All of these men clearly thought differently enough from their contemporaries to be recognized for their intellectual capabilities. It would be hard to make a list twenty times longer of equally influential right-handers. This is not to say that there is anything inherently great about being left-handed, but it does underscore the value of a different approach to problem solving. Only a few in a society get to lead it in new directions. Those few are the ones with valuable new ideas outside of the "group think" of the rest of society. Is it any surprise that so many of our intellectual leaders rise to prominence by coming at problems from a different point of view?

Left-handers used to be ostracized. When my father (a leftie) got to his new teaching assignment in Toronto, the departing headmaster let him know that he thought left-handers were essentially "uneducable." Imagine what he would have thought of the prospect of female students in this century-old boys' school.

With at least two dozen highly specialized processing areas of the brain, statistically some people will be more developed in some areas and less developed in others relative to some average. Genetics drives this "hardware" difference.

For instance, color blindness is an affectation in the brain that comes in varying degrees. Color blindness, like autism, is an affliction expressed mostly in males. It's hard to believe that anything in one's cultural upbringing is responsible for the ability of a person to see in perfect color or in some subset of the palette, or to knit the synaptic network permanently together hundreds of times faster than average.

Some genetically-driven traits of the brain persist from generation to generation. Some years ago I was asked to give a dramatic performance of my book *My Grandfathers' Clock* to a gathering of Bacons in Georgia, because of its framework of twenty-eight sequential generations of Bacon male ancestors. We quickly discerned that the line of Bacons at this reunion had split off from my own line twelve generations ago in 17th century Massachusetts. However, virtually every member of this clan could have blended seamlessly into a gathering of my first cousins on the Bacon side, and poorly into any other gathering of relatives by marriage. We Bacons are a different sort of group (as my wife will attest), with many traits unlike those of other families. We all love the woods, think in three dimensions, like making physical models and prototypes of our thoughts, make a lot of our own specialized tools (every branch of my clan has built a log splitter and its own design of a sailboat, for instance), and we like to manufacture steam and internal combustion engines. Something cognitive is being passed along in the genes of these men. We share habits and traits that I have not observed in any other group of people: say, for instance, my wife's family, or that of any of my siblings' in-laws. Since I saw the similarities mostly in the men, and they're uncommon enough that I don't see them in most other people, I'm betting that most of the traits are carried on the immutable Y chromosome all male Bacons share.

OK, so what does all this mean? I'm unique. So are you. "Come on," you say, "how different can we be, really? How much brain diversity *is* there in the world?" A lot, really.

We can make an estimate of the range of brain diversity. Assume a low number of specialized regions of the brain: say, twenty-six. Assume that each individual can have one of three rough levels of performance in each area: either average, (A), lower than average (L), or higher than average (H).[5] If we reach into the twenty-six bins of parts to assemble our sample brain, we

---

[5] Of course, there is a continuum of capabilities, but we'll just take three to illustrate the point.

have three possible types for the first region (A or L or H), and three for the second (also A or L or H). Already we're up to nine possible outcomes. One lucky person will be above average in both areas (HH), one unlucky person will be below average in both (LL), and everyone else will be in between (AL, LA, AA, AH, HA, HL, LH). Every time we add another specialized region, we get three times the previous number of outcomes. After the admittedly low number of 26 specialized regions are assembled, with only three categories of performance each, we find that we can assemble 3^{26} = 2,541,865,828,329 (or 2.5 trillion) types of brain. That's about two hundred times the number of humans (twelve billion) who have ever lived.

All of these brain types have potential value. The Olympic Games would be pretty short and simple if there was only one way to measure athletic prowess. Similarly, there are scores of ways to measure intellectual prowess. And you don't have to be gifted. Just as you don't have to play Olympic soccer to enjoy the exercise, you don't need to be superior in any part of your brain to enjoy the process of intellectual growth. And here's the other thing. In the Olympics, a single, random spectacular achievement (such as a swish shot from the opposite end of the basketball court) is of momentary value, and won't help your team in the next game. In the parallel universe of intellectual achievement, such a swish shot (like, say, the equation $E=mc^2$ from an unlikely player, a patent clerk in Switzerland) instantly becomes part of the capability of *all* members of the species, from that generation onwards. Thus, any diverse brain with an interesting new approach to a problem can light the way for all other brains, if given a chance.

Part of our current challenge is to recognize and to integrate the various types of intellect that we observe in the world. Almost everyone has something to contribute. In fact, the chances are one in thirty-eight thousand that any particular individual has no above-average mental traits, even in our simple three-grade 26-feature brain model. Similarly, it's a rare individual who is not below average in *something*. As I've said, my management knows how to channel my good traits, and to avoid loading me up with things for which I am poorly suited. Every brain gravitates to its best environment, and tries to optimize the workload to its own strengths once there.

And now some even better news. Genetic predisposition for particular intellectual talent is only that: predisposition. What really matters, as we shall see in the next section, is what patterns you load into your brain during your life. I'd rather have an educated average thinker than a gifted but uneducated superior thinker working for me, any day. As I've said, the average human brain hasn't changed in three thousand generations of humankind, and with over two trillion possible variations in a population of only six billion, we can't wait another thousand generations for a superior mind to randomly show up. I need minds working for me (and especially the one within me) with the best possible patterns, including the pattern of disciplined, inquisitive, methodical thought.

Thus, we rejoice in the diversity of talent that is available to society, and we turn our attention to education. Education still has its flaws and biases, but overall, it's the root of our evolution. Universities, steeped in male traditions stemming from the monasteries and from the royal courts, are still not fully suited to women's needs, just as some places still don't have left-handed desks for the students who need them. Lawrence Summers at Harvard and my dad's old headmaster were unprepared to see how to fit new pieces into the puzzle. They'd better learn. As we've just seen, a disproportionate number of lefties are routinely leading the free world, and soon you'll see that women are surpassing men in the number of academic degrees awarded each year in the world. The old patterns, like everything else in the parallel universe, are changing explosively. And what did the community think of Dr. Summers' gender-based philosophy? His 1997 replacement as Harvard's 28th president was a woman: Professor Drew Gilpin Faust.

WHERE DOES NURTURE COME IN?

Dr. Gregory Stock of UCLA has claimed that, based upon studies of twins,[6] only about 20% to 25% of our behavior can be attributed to any genetic predisposition, with the remaining 75% being learned. This is not surprising, considering the way that we all can function reasonably well in a common societal structure despite of the wide brain diversity claimed in the last chapter,. If you buy two computers with slightly varying chip sets, clock speed, hard drive size, etc, it is still hard to differentiate capability when they are loaded with identical programs. Granted, the one with the faster clock speed and better graphics processor may be better at gaming while the one with better sound card and higher contrast screen may be better at video streaming, but on the whole, what really determines the productivity of the computer is the quality of the software you load. The brain is no different.

Several years ago I was giving a speech in Binghamton N.Y. A nine-year-old boy approached me with his drawing of the solar system. He had correctly placed the sun in the center of the system with the planets in their proper order and distances. He'd captured the wisdom of Copernicus before he'd gotten to the sixth grade. He had drawn the orbits as ellipses and properly

---

[6] Twins studies include those raised together, apart, identical, fraternal, and even unrelated-adopted pairs.

put the sun at the focus of the ellipse. He had captured the knowledge of Kepler before he had learned any algebra. He had properly drawn the colors and relative sizes of the planets, complete with the moons of Jupiter, its bands of color, the rings of Saturn, the craters on Mercury, and he had shown the outermost planets of Neptune, Uranus (with its tilted axis) and Pluto with its tiny moon Charon. It appears that that while the young man was still at the age when girls are deemed to be "yucky," he had acquired the essences of the lifelong work and knowledge of Galileo, Clyde Tombaugh, Isaac Newton, Percival Lowell, the entire U.S. Space program, and others. He'd shown the new bodies Quaror and Sedna. Just out of curiosity, I asked him if he understood what $E=mc^2$ meant, and he responded without a pause that it meant that a lot of energy could be made out of matter, and that this was the process that ran the sun. It appears that the young man was pretty well up to speed for a nine-year-old, and ready for another three years of lessons before heading off to high school.

    Of course, he didn't have to derive all this information on his own. He had the advantage of Gutenberg's press, Marconi's radio, Farnborough's television, Bell's telephone, and Shockley's transistor to help shoehorn all of these valuable patterns of thought into his young mind. With these foundations, he'll be ready to tackle the next part of the puzzle. Who knows…maybe he'll solve string theory.

Why did that boy's teachers and parents choose to load up his neural net with a sun-centered solar system, elliptical orbits, and relativity? Why did they choose these models rather than mysticism, alchemy, and earth-centered crystal celestial spheres? It's simple. The older ideas are extinct, and better ones have evolved. Those new ideas and millions of others are fighting for resources in every human brain. Fifty thousand CD ROMs are fighting to get into every young skull. Only the best ideas win.

In his book *The Selfish Gene*, Richard Dawkins in 1976 identified the key way to understand our evolution as a species, in terms of this competition for brain energy. Dawkins identified that human thoughts are *self-replicating patterns* stored as neural synapses, just as genes are self-replicating patterns of information stored as base pairs along a DNA molecule. *Replication* is the key indicator of whether anything will survive in preference to random chaotic other outcomes. Good ideas replicate, and bad ideas die.

Ideas, like genes, are little packets of information: pieces of stored complexity and order that have developed the amazing ability to make copies of themselves. Ideas are copied every time a thought is transferred from one mind to another. They are information that is replicated from one human mind to another.

Remember? The mind stores sixty-seven thousand times the information of the genome. That's sixty-seven thousand times the information that must be coded in each generation, starting from scratch. Clearly, information and thought patterns that work better will be preferentially passed on to the next brain.

Self replication is what evolution is all about. It helps to recast the problem of reproduction in terms of what genes do, not what the host body does. (Consider that a chicken is an egg's most efficient way of producing another egg...) Genes and ideas are both replicators.

Dawkins and thousands of followers have termed the nuggets of intellectual information, theories, and models to be replicating entities called *memes*, and the idea of the meme (the "meme meme," so to speak) is now so powerful that the word appears on over thirty-five million web pages, only thirty years after the concept was first coined. The meme is a really great idea that fits what we observe, so brains all over the world are suddenly using the concept of a meme to understand many features of human learning and problem solving.

Just like genes, memes replicate pretty fast, once they prove their superiority over other competitors. In testable scientific matters, this process is almost instantaneous. Who teaches the death of the dinosaurs by creeping ice age anymore, only thirty years after one original brain[7] proposed the idea (and the showed the evidence) that an asteroid hit the Earth 65 million years ago?

---

[7] The brain belonged to Walter Alvarez.

In matters of religious and political beliefs (where the whole foundation of most faiths and political groups is to preserve ancient beliefs and existing structures), the evolution of memes is slow, contentious, and painful. The Vatican only forgave Galileo for his heresy of claiming that the Earth revolved around the sun four hundred years later, in 1992. At least he lived out his days. Bruno was burned at the stake for making the same claim.

Once a replicator (gene, meme, or some hypothetical other form) comes into being, it must make copies of itself, or be washed away with time. Dawkins succinctly captured the three basic traits that lead to preferred survival of a replicator: *Fidelity*, *Fecundity*, and *Longevity*. These are all true of genes and memes.

Longevity: the longer a replicator exists, the better. It becomes part of the local ecology, consuming resources at the expense of other potential competitors. All life forms try to hang on as long as practical. So do ideas. Ideas caused people to ban books and to burn Bruno.

Fecundity: The more copies of itself a replicator can make, the better chance it has of surviving through to the next generation. Frogs make tadpoles. Memes make speeches, books, and political posters.

Fidelity: Generally, it pays to make exact (or nearly exact) copies of the replicator. Slight variations (mutations) are useful only when there is significant competition for resources, only if the modifications are slight, and only if they occur in just a small fraction of the copies. If some of the many copies are going to die anyway for lack of resources, it pays to tinker slightly with a few of them. At worst, the tinkered ones fare worse than their peers, and become part of the fraction that was going to die anyway. However, the few variations that actually fit the environment better will quickly succeed, and their high-fidelity copies will quickly displace the high-fidelity copies of the un-modified variety as the competition for resources continues. A replicator needs a lot of very good copies, and a very few fairly good ones.

The resource for which each meme is competing is the attention of (and energy expended in) the human mind. Actually, it succeeds by being *efficient* in its use of mental energy. If the model you teach to your child explains everything she observes in the world, and predicts certain true things, then it's a valuable pattern. The mind spends its energy establishing new links between neurons, attempting to match new complex patterns in the real world with simple, established patterns in the brain. If it's an easy match (a stoplight at a new intersection, or a lunar eclipse, for instance), we spend very little energy, match a behavior to the stimulus, and get on with our lives. Whenever no match is evident, the brain spends energy trying to find enough pieces to explain the observed phenomenon. A lunar eclipse is stressful until you assign this a pattern of being an act of God or of orbital mechanics. If a green light flashes while the red light remains lit, some of us start problem solving. Canadians have a pattern that says: you have right-of-way to turn left *now*. This is useful if you have a mental pattern for it, and stressful if you don't.

HOW POWERFUL IS AN IDEA?

The federal government can throw someone in prison for a very long time for a variety of reasons. It can only execute a person for a few of them. Other than a few really ugly crimes against innocent bystanders, especially children, all of the big penalties are associated with interfering with decision-making and fact-finding. One class of these crimes is assassination of a federal elected official, law officer, judge, prosecutor, or witness. Another is treason, espionage, or the destruction of infrastructure necessary to conduct government business.

In the former cases, the perpetrator has of course denied the official his or her life, but he has also interrupted the flow of policy and ideas that are to be shared by the population at large. In the case of treason, the perpetrator has released valuable information that our government knew, but another did not. In both classes of crime, you can be sent to death for compromising information and decision making.

Many religions have a death penalty for beliefs counter to the ideas central to the faith. This form of religious treason is called heresy. The attacks of 9/11/2001 were all about ideals that

America did not embrace, that were central to a fundamentalist Islamic sect. Sunnis and Shiites are currently battling out matters of religious difference, with fatalities mounting daily. The Spanish Inquisition of 1481 to 1492 murdered thirty-one thousand "infidels" for veering from the Catholic faith, and the Crusades were amongst the bloodiest escapades in human history, all fought over matters of belief. Quakers were executed by the Congregationalist Protestants in New England in the late 1600s (right up there with witches). Recently, several people lost their lives at a meat market in Indonesia because pork was sold there.

Note that during the French Reign of Terror of 1793 to 1794, the Russian revolution of 1917 and in the Third Reich in the 1930s and 40s, there was a systematic round-up, quarantine (and usually, execution) of the intelligentsia—those members of the larger society who generally lived in the world of creative thought, and who propagated new ideas. Independent creative thought was frowned upon when uniform social ideology was being rolled out, especially when such ideology included centralized policy-making. It wasn't so much the material assets accumulated by the intelligentsia, but rather their access to the hearts and minds of the populace that were the targets of the rising powers.

In the case of the Nazis (and much later in their own regime, of the Soviets), there were powerful efforts to substitute propaganda for intelligent debate, to assure uniformity of thought.

We see remnants of propaganda today in the "spin-doctors" of modern American politics.

Although American politics is becoming more distasteful by the decade, there are a few features of American society that have traditionally helped it to be strong, and it is interesting to note how the evolution of *ideas* is central to all of these facets.

1) America is a democracy. In principle, the candidate with what voters perceive to be the better ideas wins.
2) Our elected officials are term-limited. We force the refreshing of intellectual leadership and ideas after at most eight years in the presidency, even if we've liked the incumbent. We reaffirm our leaders' ideas, or replace them.
3) We're tri-cameral. Our executive, legislative, and judiciary branches all keep each other in balance. No single branch of the government can perpetuate itself or its policies. Ideas are weighed and balanced.
4) We're capitalist. We let individual ideas compete in the marketplace. The best ideas in product or marketing usually get rewarded.
5) We're free-market. Although there are some exceptions, we allow and encourage foreign competition for American dollars. This allows other cultures and value systems to show us ideas of how to improve, even at the expense of American jobs.

6) We protect ideas. The U.S. patent office is the most active in the world, issuing a seventeen-year patent on average every thirty-two seconds of every working day, and across town, copyrights are recorded at an even higher rate. Those copyrights last for seventy years.
7) We are inclusive. Our constitution prohibits discrimination based upon almost any measure other than intellectual ability. Race, religion, gender, national origin, age, and physical ability are all equally protected under the law for most jobs in the U.S., and the only exception for that list is physical ability in jobs that actually demand it. The only thing where an employer is always free to discriminate is in intellectual ability—the ability to create ideas.

Although we are far from being perfect in our implementation of any of these facets, we try to adhere to the principles. Every one of the principles is, at its root, designed to reward, defend, and encourage better ideas. It is interesting to note that in the last fifty years, ninety-five of the world's governments have migrated to free-market, term-limited capitalistic democracies (117 of 192 countries in 2000 versus only 22 in 1950). The ideas of how to promote ideas are spreading. The many variations on democratic free-market capitalism may soon evolve into even stronger and more affluent social structures than America's.

The Four Factors

MULTIPLE FACTORS

I see several factors affecting the rate at which humans acquire knowledge. These factors are nearly completely independent, meaning that they multiply each other. I believe that the factors, in decreasing importance, are as follows:
1) An increased availability and variety of educated brains.
2) Unprecedented cross-pollination of ideas between previously segmented areas of science, as a unified and consistent model emerges in critical mass of correlated thought across all cultures and languages,
3) Silicon-based tools of computing and communications, which now allow alternate, faster pathways and environments for memes to evolve, and
4) Increased understanding and enhancement of the mechanics of thinking and of creative thought, including psychopharmacology, cognitive science (with five-dimensional modeling of the human brain at work), and the study of memes as replicators.

Some of these major factors have sub-factors and multiple additional dimensions. Multiplied together, their influence on the rate of human understanding is enormous and is growing explosively.

FACTOR ONE:
THE AVAILABILITY OF BRAINPOWER

1a) THE EDUCATION BANG

In 1900 Cambridge University produced one thousand bachelors' degrees per year, and the population of the world was about one billion. In the first decade of our new century Cambridge is graduating over six thousand graduates per year. But that's just one university. Scores of new ones have originated since then, handling not only the population growth, but the increase in the fraction of the population that is receiving degrees.

Surely all those six thousand brains from Cambridge are producing intellectual capital at a higher rate than their one thousand brains could do a hundred years ago.

The population of the planet is expected to grow to twelve and one half billion people in the next hundred years. If the growth of society's collective intellect is provided by just the fraction of the population in all academic and technological pursuits, then does it not make sense that with twice as many brains then as in today's society, that even a common, *constant* fraction involved in new creative output should produce approximately twice as much creative output in total in the society that we see today? Sure.

We can get some idea of the accumulated output all of these deep thinkers by such technological indicators as the gains in our agricultural productivity and gross national product (GNP). We can even directly measure the raw production rate of good ideas by counting patents.

One hundred years ago we created twenty thousand patents per year. Today the U.S. Patent Office issues a new patent every forty seconds, for a total of over one hundred eighty thousand per year, or nine times our previous rate. Inventions are pouring out of Western minds at a rate faster than any one mind can comprehend.

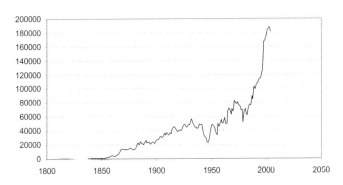

Figure 5: Patents per year awarded in Untied States vs. time

We know that the population of the U.S. is growing, and that we have approximately three hundred million people within our borders now. Is our patent explosion just a reflection of the population explosion? No, there's something more going on. It used to be true that patents and gross national product grew only slightly faster than the population: from 1900 through 1979, we quadrupled our patent production while we tripled our population. Since 1979, however, the population has grown by only twenty-five percent, but our patent rate has more than doubled. In that time we've increased our *per capita* patent productivity to *two hundred seventy-eight percent* of our previous rate, from a 1979 low of 2.3 patents to 6.4 patents per ten thousand people now.

Thus we're more productive as a society than logic would dictate if we had a constant fraction of us contributing the new ideas. Are we growing *better* brains? Yes and no. Genetically, it's the same old brain. We have an increased number of brains, but we are also dedicating a larger and more diverse fraction of that brain pool to solving our species' biological, societal, global-scale problems, and a smaller fraction to solving local, personal subsistence, home-scale problems. Each new step forward in the global arena is a step forward for all of us, and those steps are coming at least nine times faster than they did a century ago.

Let's look at agriculture as a major example. Approximately one thousand years ago, human society reached the limits of what the arable land of the planet could support, based upon the limits of human and domesticated animals' physical strength and endurance. We could only grow so much food, and with periodic swings of feast or famine, war or plague, or of new lands opening up, the population of our planet stabilized at approximately one billion people for nearly a thousand years. With the rise of the Industrial Revolution, the population started to increase again, based not upon our finding of new lands, but upon our technological base that has allowed agricultural productivity to support a six-fold increase in the population with a proportionately smaller fraction of that population in economic or physical hardship. That gave us the increased potential supply of brains, but technological advances changed the ratios, too.

Consider that one hundred fifty years ago over sixty percent of the population of the United States was involved in producing the food supply. Today we need only two percent of the American work force to produce more food than the nation can consume. It's more food than is needed for economic viability. We export vast quantities of it. We subsidize our farmers not to grow crops so that they maintain a strong economic foundation. It is not that the citizens of America have evolved to need less food. No, we are awash in an abundance of food, and obesity is growing to be one of the top national health concerns. The only thing that has changed in the last one hundred fifty years is the *technology* of agriculture. We've adjusted our means to harvest, to store, and to distribute food, and have also adjusted the biological makeup of the crops themselves. Ninety-four percent of the former workforce has left the agricultural sector, generally for more academic pursuits.

1b) THE DIVERSITY BANG

As we have discussed, women's brains are different than those of men, and it is apparent that most fields of study are benefiting from a larger diversity of thought patterns and mental processes by including women. If we believe from our earlier discussion that there are subtle genetic differences in mental makeup, it should be encouraging news that virtually every branch of humanity is growing in the total and in the fraction of degrees awarded.[8]

My grandmother Grace Dunlap was the second woman hired into the National Advisory Council on Aeronautics (NACA) in 1920, following her sorority sister Pearl Young, each of whom had a master's degree from University of North Dakota. These two women pioneers in aviation technology took their jobs in the same year that the U.S. Congress granted them the right to vote. At the time, only twenty percent of American women had any sort of paying job, and few of those jobs were in the creative core of the business world. Today, seventy-eight percent of adult women have paid jobs, and at the National Aeronautics and Space Administration (the agency that evolved out of my grandmother's NACA), approximately forty percent of the technical work force are women.

---

[8] Only Native Americans are not increasing their statistical proportion of representation in the academic ranks.

In 1900, less than one woman in thirty had a college education. By 1973, it was one in seven. As of 2006, the number is approaching one in four. Women receive forty-five percent of all the degrees awarded in all fields, and women outnumber men for the first time in history in several fields, including life sciences, education, the humanities, and the social sciences. They still lag behind men in achieving degrees in engineering and in the physical sciences (seventeen and twenty-eight percent of the degrees awarded, respectively, in 2003).[9] However, the universal trend in all disciplines is upwards. Women are advancing on every academic front, at all degree levels.

One might question whether the addition of women into the workplace has added to the net brainpower, or simply raised the competition for the equivalent number of positions. If this latter statement were true, one might expect to see approximately forty percent unemployment in the United States based upon forty percent of the previously male-dominated jobs going to someone else. Clearly this is not the case. Unemployment in America has declined to as low as four percent while women have more than quadrupled their percentage of the workforce.

Women are corporate executives, bankers, engineers, doctors, astronauts, and politicians. In

---

[9] Data from NSF/NIH/USED/NEH/USDA/NASA, 2003 Survey of Earned Doctorates.

only the last generation women have been elected as the prime ministers of Finland, Germany, Great Britain, India, Ireland, Israel, Liberia, Pakistan, the Philippines, and New Zealand. Until 1981 there had been a total of five women who served as a U.S. state governor—all since 1925, and most succeeding their deceased or impeached husbands. Since then, twenty-two women have been elected as governor, some in head-to-head races with other women.

Officially, women are fully engaged in every part of American business, science, and technology. However, there is still more room to grow before true balance is achieved. Look around your own place of business. What fraction of your workforce is women? When might it be fifty percent?

As I've said, within NASA women and are approximately forty percent of the workforce (both in the civil service and in our contractors). This makes some statistical sense. Nearly eighty percent of the female population (fifty percent of the total population) is in the workforce, for an expected fraction of forty percent. It's good that NASA actually has even a statistical average, because of the small fraction of scientific and technical degrees awarded to females (so far…). However, I am keenly aware from my world-wide visits and lectures that in many other corporations women are a smaller minority. I personally believe that the more creative and dynamic the setting, the more likely you are to attract and keep a diversity of brains, and that one is likely to see more resistance to minority views and talents in more entrenched industries or academic institutions, where change is not perceived to be as imperative and traditions are embraced more dogmatically.

There is an advantage to having a diversity of opinions and problem-solving approaches on any team, and it seems clear to me that the addition of women into traditionally male occupations, particularly the intellectual ones, can only foster a wider range of options and a better solution to most of the problems encountered by the team.

Note that I've been discussing the rise in the population of educated women. Not white women. Women. All women. Meanwhile, the formerly dominant white males are going to college in larger numbers, too. So are minority males. Nearly all ethnic groups in America are going to college in numbers quadruple, quintuple, or even tenfold their previous rates. Because there are so many more degreed professionals joining the workforce than degreed professionals leaving it, American society is currently blessed with eleven times as many living college graduates as it had only a generation ago.

One hundred years ago, five percent of the U.S. population achieved a bachelor's degree or higher. That was adequate when so much of our workforce was needed in agriculture. Today, many of the fifty-eight percent of the population no longer necessary for food production are turning to more creative pursuits in the office world. Even of those 2% of society who remain in agriculture, many are in the intellectual end of the business, administering the business of farming, soil science, plant genetics, advanced weather forecasting, etc, while machines do most of the labor. All of these facets require higher education.

Figure 6 illustrates the rise since 1940 of the percentage of different demographic groups achieving bachelors' degrees in the United States. Note that *every* demographic group is rising, most by almost a factor of ten. In the white majority, bachelor degrees are being attained by over forty percent of the population. .

This trend is not just in America, and it's not just bachelors' degrees, either. U.S. doctoral diploma awards have risen by a factor of 2.6 in the last twenty years, keeping pace with Japan at 2.3 times the doctoral achievement rate of two decades ago, and India, which has seen an exponential growth in all degree levels and fields since the mid 20th century. Women, once only a tiny fraction of the degreed population, now outnumber men in most academic fields (even with a tenfold increase in white males achieving the bachelor's level), and this is especially reflected in the rate of doctoral degrees. Although women do not yet outnumber men in science and technology, that disparity is rapidly shrinking.

White European males no longer monopolize the creation and flow of ideas in the major economies of the world. (Rembrandt's painting *The Dutch Masters* comes to mind, as an example of the way things used to be.) Not only is the creative core growing to include significantly higher raw numbers of degreed workers in our new knowledge-based economy (approaching a factor of ten, compared to a few decades ago), it is embracing a much wider diversity of cultures and, presumably, natural intellectual strengths. Moreover, the global economy assures true diversity. Any homogeneity we might have had across ethnic groups within our borders (due to common language, government, tax structure, and other averaged cultural values) is overwhelmed by the international influences on every part of our new society, where cultures are even more distinct and diverse from our own.

We explored in a previous chapter the idea that diversity in brains is good, because it leads to a variety of strengths and perspectives: essentially a richer variety of environments where emerging memes can better sprout. We also considered that once an idea has evolved, it empowers everyone who adopts it. Imagine what lies ahead as the educated fraction of society is eleven times its previous size, and dozens of times more diverse.

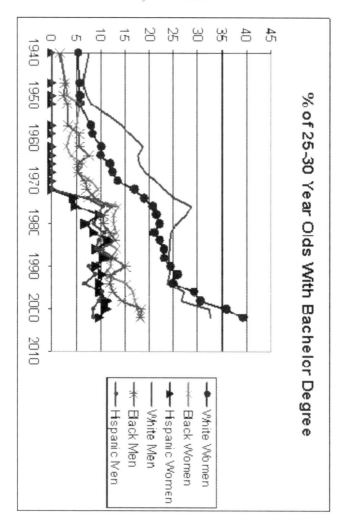

Figure 6: Percentage of 25-30 year olds with Bachelor degrees, by ethnicity and gender.

1c) LIFE EXPECTANCY

A brain can only contribute to society's knowledge once its owner has survived to maturity, and can only continue to contribute as long as the brain's owner is alive. The great news of the past century is that both of these criteria are now greatly enhanced over any period in human history. A brain's productive life expectancy is growing.

The most dramatic gains have been achieved in the survival rate of infants and of young mothers (who now die at a tiny fraction of the previous (1800) rate of twenty percent death by childbirth), leading to a vastly larger fraction of the population reaching old age. With the elimination of a large number of early deaths, the life expectancy at birth now predicts that most of us reach the general life expectancy formerly forecast for the few people who had already successfully reached the age of fifty. Based strictly on the availability of mature brains, we observe twice the available problem-solving capability per birth as we had only three generations ago. That partly explains the population growth. But for a meme, real longevity counts in adult life, when problem solving and experience are at their zenith.

We have been making slow, steady progress in increasing human longevity amongst those who survive past adolescence. For the past two decades, we've been raising the average lifespan of Americans by approximately forty days per year, or more than two years during the last twenty. That means that the U.S. population has increased the mental production of its adult workforce by over five hundred million individual work years during the last quarter of their lifetimes, just by giving educated adults more time to ponder things. This is the equivalent productive output of all the courts of Europe throughout the Renaissance, or of Europeans in North America for the first hundred years of colonial life, or more effort than it took to build the Australian Navy—just for living longer.

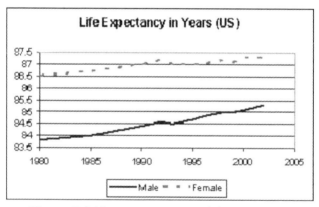

Figure 7: Life expectancy on adults over 60 in the United States, vs. time.

I don't expect these trends to continue. I expect them to get even better, and that we will live longer, on average, than current tends predict. Perhaps there is a physical limit to our average age, but I'm sure that we're far from reaching it.

Why? Today, the largest and wealthiest population of people in history is approaching its declining years. Those very wealthy people are busy investing, and also busy trying to survive longer and more comfortably. They are the largest single voting block democracy (or capitalism) has ever seen. Companies that cater to quality of geriatric life are blossoming and fiercely competing for these customers and for their hundreds of billions of dollars' capital. Just as one example, note that heart surgeries on humans have gone from none in 1940 to one million, seven hundred fifty thousand *per year* in the U.S. alone. At this rate over the current U.S. population, one in every two Americans may ultimately have elective heart surgery to improve their quality and length of life.

Today we have an explosive growth in theoretical understanding of biochemistry and gerontology, and increasing non-invasive diagnosis and corrective surgery. Will you be surprised if we add another four years of average quality life to our population before we baby boomers die, instead of just two more? That would be one half billion intellectual extra work-years out of the current American population beyond the half-billion increase we currently expect (or two *more* Australian Navies!).

FACTOR 2:
CRITICAL MASS

When geologist Walter Alvarez proposed that the dinosaurs had died from the impact of a large asteroid with the earth (and from the global climate catastrophe that followed), the idea was slow to spread. When Milislav Demerec proposed that biological mutations and cancers were at least partially caused by subatomic particles and high energy photons disrupting the molecular chemistry of DNA, there was similar foot-dragging. It was a physicist, Max Delbrük, who won the Nobel Prize (1969) for his study of the physics of how radiation actually created a mutation.

These days, it is far more common to cross technical disciplines, and similar fundamental truths are being uncovered. Geologists, life scientists, physicists, astronomers, engineers, climatologists, and other intellectuals have increasingly recognized the unity of science, and have seen that not only are there are scores of ways that we can find truth, but that it is not truth until it is consistent in scores of disciplines. Like neutrons cascading throughout a critical mass of uranium, intellectual energy is streaming out of previously isolated lumps to trigger vigorous debate and unlock new energy in distant areas. Those debates look to unify local theories with previously unrelated models in other fields, to assure that the models not only explain the local field of study, but are consistent with other findings and fields. Such multiply-consistent models are more likely to be true than models that do not possess these key features. As scientists cross fields of study, it becomes more apparent that the models are indeed consistent. Consider, for example, the following chain of ways to explain the world, which is consistent, testable, and repeatable.

Electric forces and quantum mechanics *can precisely explain and are consistent with* how an atom's electrons populate stable shells around the nucleus. These same theories *can precisely explain and are consistent with* the way atoms bond into molecules, and define the electric potential around such molecules, that determines how they attract or repel water and other molecules. These electric attractions *can precisely explain and are consistent with* how DNA makes copies of itself, and how proteins and enzymes work. They even *can precisely explain and are consistent with* mutations, when coupled with understanding of nuclear particles (Delbrük's Nobel Prize). Mutations and DNA copying *can explain and are consistent with* the extent and variety of plants and animals we see on the planet. The adaptation of animals to their environment through mutation *can explain and is consistent with* the observation that one animal (the human) has mutated/evolved to fill an ecological niche by using neurons more than other animals. The conductivity of certain combinations of chemicals *can explain and is consistent with* how neurons fire or are inhibited. The firing and inhibition of neurons *can explain and is consistent with* how patterns can be recognized and matched, and how new information can be synthesized. Such pattern transmission, recognition, and information synthesis is evident in your contemplation of this printed paragraph that begins: "Electric forces and quantum mechanics can explain…"

That initial single theory of charged particle interactions is consistent with the chain that follows. One might challenge certain links. One can believe that some other explanation could also explain and be consistent with what we observe. However, the global connectedness is key.

We can make a tenuous case for almost any causal link you can imagine, just as you can probably hammer any two jigsaw puzzle pieces together, even when they don't really fit. However, there's only one way to put a jigsaw puzzle together correctly. If pieces don't fit, we soon spot that there's a problem as we try to build out the picture.

If science is a jigsaw puzzle, we are learning that it is *all* connected, and that the unsolved regions can be approached from any one of a number of directions. That's powerful.

Like a jigsaw puzzle, the solution is being generated faster as more of the puzzle is solved. Folks working on the borders see where the folks working on the horizon, sky, and foreground fit in, and vice-versa. The holes in the puzzle are easier to spot and to fill. The great goal is to unify the different regions, and as each region builds, the interfaces and unions with other regions become more evident, and more a cause for collaboration.

Like a nuclear pile, the energy streaming out of one area is triggering the energy in other areas. For the first time in history, it all *does* appear to fit.

All we have to do is look at the whole puzzle and share our work with *all* the other billions of puzzle-solvers. The fact that there are so many of them and that there is so much to know—so many pieces of the universe-sized puzzle already in place—would ordinarily have been a problem for our little brains, if we'd had to share information in the traditional ways. That problem just got fixed, however with Factor 3: the Silicon Bang.

FACTOR 3:
THE SILICON BANG

In the opening chapter, we explored some of the effects of modern computing on our understanding of the universe, and marveled at its pace. Now that we understand the power of memes as replicators that guide our evolution as a species, let's return to computing to explain *why* silicon logic is advancing so fast.

Think about it. We humans evolve by our ideas. We evolve by gathering, sifting, sorting, storing, correlating, and sharing information. Our ideas emerge back into the real universe in the form of our technology. Our technology allows us to work with the forces of nature (including biology) to reduce our expenditure of physical energy, while improving the way we eat, drink, live, transport ourselves, and build physical shelter. Our technology has shown signs of significant growth *every time* we improved our ability to gather, sift, sort, store, correlate, or communicate our information, such as with the establishment of language, writing, the printing press, the scientific method, or ships in every port-of-call.

After a million years of human ascent from *Homo Erectus*, our information tools have "recently" (i.e., since the Renaissance) allowed us to double our technology every generation, until this one. That's when we invented a technology that does what we uniquely do to evolve. It sifts.

It sorts. It stores. It correlates. It communicates. Unlike any other technology in our last million years, we have channeled our resources to double *this* technology every five hundred forty days. This technology enables geometric growth in the essence of our evolution, and consciously or subconsciously, society is putting as much relative energy into running its silicon brains as our bodies put into our neural brains.

Silicon logic affects us in several ways, so let's examine a few of the key ways from a meme's-eye view.

3a) COMPUTING

It's been written that there are 10 types of people in the world: those who think in binary numbers, and those who don't. (If you don't get this joke, don't worry. It just means you're not a computer geek. It might help if I suggest that there are 00000010 types of people in the world…)

A computer, at its simplest, manipulates *binary* (yes/no) numbers. Because our brains have a thousand possible neural channels to follow for each neural input, people don't normally think in the cumbersome (but very reliable and repeatable) yes/no binary logic. We're more like a base-*one thousand* computer, as opposed to a base-*two* computer. Thus in the early years there was a general incompatibility of computer "thought" with human thought. Computers just didn't understand us, no matter how much we yelled at them. There were enough useful things to do (especially crunching mathematical equations tirelessly) that only a few people put the necessary effort into learning to translate their thoughts into the cryptic bits and bytes that would allow the silicon to do its very repetitive work. Few of us needed the same simple mental job done over and over again, or to several hundred decimal places of accuracy, so the computer was relegated to NASA and the military and some back rooms at universities.

A few adventurers started to put sensors and effectors respectively on the input and output side of computers, so that they could, for instance, learn from the sensors where the rocket was pointing, compare the input number to one that the computer had calculated to be the right place to be, and then subtract the two answers and feed an appropriate signal to the rocket engine to point the rocket exactly back on course. Still not a mass-market kind of thing, and only a few of us got to use this kind of capability (in this case, to go to and from the moon) even with a half-ton computer doing this very sophisticated work for us.

Gradually, however, Moore's Law was giving the computer more capability to process its binary bits. It was able to store and manipulate more data, so we could use lots of bits to represent text (we needed eight bits for each letter or numeral) and keep our data correlated. We got good at teaching the computer to sift through lots of data, looking for patterns, and we got it to be better at using bits to help us communicate with it, with cursors for input, and colors and graphs and even speakers for output.

When I was a graduate student I started to have my thesis typed in 1980. It took awhile, but that's another story. As it turns out, I was the last graduate student in the engineering college to have his thesis hand-typed. The guy behind me had started to experiment with the text processing capability added to the department's supercomputer. (The supercomputer, a CDC 6600, did the implosion calculations on our thermonuclear energy project. It has since been far surpassed in processing capability by today's PlayStation II®). He discovered that while the original typing didn't go any faster, the *revisions* went like lightning!

Text processing blossomed into the fore of computer applications, Xerox made some astounding leaps forward in the field, the mouse was invented, and the personal computer showed up shortly thereafter. Still, they were cumbersome to use, and very hard to program. However, the binary bits kept coming cheaper and cheaper, and we used more and more of them to let the computer be a little more imprecise—a little sloppier—a little more multidimensional. In short, computers started to evolve a much more *human* way of interacting with us.

I worked for a long while in the field of artificial intelligence, which makes a computer appear as though it's thinking like a human. Deep down inside, the computers were processing just as fast (or faster) than ever before in their binary way, but at the interface, they were giving us black text on white pages, beeping if we made a spelling mistake, and most importantly, meeting us on our terms. They got good at understanding and synthesizing speech, and at recognizing printed and handwritten text. They knew that we were messy and disorganized, and let us work on several documents or spreadsheets at once, scattered about a desktop, on top of each other. All the time, they kept weaving into our lives, into our toys, our automobiles, our elevators, our banks, our grocery checkout stands, and other places. Wherever we have found a way to pull computers into a facet of our lives, we've done it. For 99.9% of us, we've done it in the last twenty years. We've achieved what Mark Weiser has termed "Ubiquitous Computing."

3b) UBIQUITOUS DATA

Memory is free. That's a pretty good approximation, anyway, and has been true since the early days of ubiquitous PCs. In 1956, a megabyte of silicon memory cost seven thousand dollars. In early 2006 I saw an advertisement for a 250 Gigabyte drive for $49 U.S. dollars (after rebate), or twenty thousandths of a cent per megabyte (twenty *billionths* of a cent per byte!). This represents a drop in price per bit by a factor of thirty five million in fifty years.

As our parallel universe expands, it is filled with bits that represent the physical world. Some of those bits are archived information, like books, tax returns, and census records. Some of the information is temporal, like the value of a particular stock, or the ocean surface temperature and wave height five kilometers due west of Key West at 6AM this morning. We store it all. At twenty thousandths of a cent per megabyte (and falling), how can we afford not to?

You may note (if you do the math) that memory *per unit cost* is not keeping up with Moore's Law, which technically is the rate of increase of how much computing you can accomplish on a given amount of silicon, and not what it costs. The latter rate (doubling of performance per unit cost) has been holding steady at the rate of once every twenty-one months, and is known as Roberts' Law.

In the opening chapter, we talked about the emergence of radio frequency identification tags (RFIDs) on consumer goods. The key enabler of this technology is the combination of techniques that now allow manufacturers to print semiconductors, conductors, insulators, and battery chemicals in layers on any substrate, like, for instance, the paper book you're holding. There is a huge push underway to replace (or augment) the barcode on the back cover of each individual copy of the book with a printed transponder and data/logic circuit. This electronic identifier can interact *at a distance* with the inventory and security systems of the bookstore. The target cost is less than one cent per label (comparable with the integrated cost of printing and using a barcode today), and it is expected that you will see these printed computer/transponder circuits on every can of soup and paperback book by 2014.

The radio transponder part of the circuit has been around for awhile. It's basically a laminated foil antenna with a foil capacitor, which you see in expensive books before they leave the shelves, as security tokens. The hard part was getting the data into the transponder, so that it doesn't just respond "something is here," but instead responds "The Parallel Bang, ISBN 0-9708319-3-5, By Jack Bacon ©2008 2nd printing, copy #1073 is here." Moreover, most store-owners would like such labels to be *programmable* to compute helpful bits like: "move me to the discount table on February 17."

Seems like science fiction, doesn't it? Actually, it's already here. For some time we have been able to add a tiny chip of silicon with very simple circuits to printed transponders, at a cost suitable for highly valuable or sensitive items, but not for general consumer products. The real science fiction dream is coming true now, though. Printable inks that dry to semiconductors, insulators, and conductors and simple lamination techniques already commercially produce (on plastic or paper) specialized circuits that include solar cells, batteries, inductors, resistors, capacitors, semiconductor logic, clocks, light emitting diode displays, and even loudspeakers. The trick is to merge all these functions into more ubiquitous and inexpensive circuits. The RFIDs are the ten-billion-dollar business opportunity that is spurring the development along, but more applications are coming. There are already pharmacy pill blister packs that record the time and dosage that you take, and prompt you for your next dose. Variants allow you to use another cheap disposable printed circuit to interact with your home computer, downloading your pill package history to your doctor's office computer. Interactive paper posters and placemats have been developed, and packaging will soon include sensors to tell you if the meat, fish, or milk has gone bad, no matter what the printed expiration date (a guess, at best) might say.

Information about the world is about to be interactive, and everywhere. Think of it as the multiplication of effects of the printing press and of the computer—two of the most significant enhancements to intellectual evolution in thirteen billion years, coming together this decade. It has been said that one magazine printing press can currently produce more semiconductor logic in an hour than the world can produce on silicon in a whole year. Imagine where it will be in a decade, with thousands of printing presses applying this technology to most of our printed items.

3c) COMMUNICATIONS

Humans now communicate nearly five exabytes of information to each other (or at least, to each other's computers) every year. That's five million terabytes per year, or more information *each year* than has been communicated through every spoken utterance of every single human *combined* over two million years.

The capacity and the price per bit for communications between computers (and people) are accelerating a lot faster than their computing equivalent rates inside the computers under Moore's or Roberts' Laws. The number of communications gigabytes per dollar has been doubling every twelve months since 1995. Router switch speed has been doubling every six months, and web traffic has also been doubling every six months. That's great if you have a computer. What if you're one of the eighty-five percent of the world population that doesn't?

That's where Nicholas Negroponte and the One Laptop Per Child (OLPC) foundation fit in. Negroponte's team announced in January 2005 that—under a different model of how to use the doubling speeds of silicon computing and communications—it would be possible *today* to build small, hand-powered wirelessly-networked laptop computers for less than the cost of the books that students needed for a full education in the third world—about $100. With such a device, manufactured in high volume and minimal (10%) profit for some large-enough manufacturer, millions and soon *billions* of children would have access to the world body of information, to video programming, radio, telephone…in short, to all the information advantages of the knowledge-based, industrial world. Ten months after the announcement of the concept, the prototype device was shown to the United Nations. Two months later, with millions of charitable seed dollars each from Google, AMD, Red Hat, News Corporation, Nortel, and Brightstar, the initial manufacturer Quanta Computer was selected and factory tooling commenced, with a firm commitment to the target price and volume ($100 each, delivered, with software, to six million children in seven test countries). No one in history has ever produced so much information storage and communications capacity in so short a time. Talk about a meme evolving rapidly! Imagine what this one idea will do to the growth of human understanding, when nearly six billion more minds are added to the few who already have silicon computing assets.

Meanwhile, citizens of the industrialized world now use the Internet Protocol to carry voice, supplanting much of the land-line traffic that was formerly dedicated solely to voice. Digital data outpaced voice communications only in 1999. It now so dominates the worldwide data traffic that voice, and now video, at just a few percent of total traffic, are carried along just like the other types of information on the Internet. It's all just information, so why separate it, and especially why pay for a separate channel? I routinely call friends in New Zealand or Wales for no added cost, by using my Voice over Internet Protocol (VoIP) software and a computer headset.

Figure 8. The beta test system as deployed in 2007 by the One Laptop Per Child (OLPC) foundation. The first of six million of these were shipped in November, with subsequent price drops and volume increases expected in the coming years, to the point when every child on Earth will have his or her own machine

3d) INDEXING: THE KEY TO THE WEB

We may have ubiquitous information, but what happens when the piece you need happens to be in Germany (or worse, in Germany and in German). In my teen years, if I wanted such a document, I would need to first know it existed, second go to a library and find a library scientist who could locate it for me though an overseas linkup to specialized databases in Germany, third arrange for shipment of (and payment for) a copy of the document, fourth, find a translator, and fifth, read it. These days, it's a bit simpler.

First, I type some simple relevant words into my web browser. The paper's web link and short abstract magically appear in milliseconds. Second, I read the English-language translation of the target paper if it's available, or any of dozens of similar references in English (also retrieved by my browser), which may reflect the same findings. If it happens to show up only in German, I plug it into any of several free German-to-English translation packages to get an instant rough translation, and *only* if I really need to go deeper, then I recruit a human translator. Generally, though, with over ten billion English-language web pages out there, I have the information I need in seconds, not weeks. I don't even have to leave my desk. I can do the same in a commercial aircraft in flight, if I choose to, or at the local coffeehouse.

Ten billion pages and growing. Just about everything you want to know about anything is out there, and it's amazing to see the frequency with which certain words show up. This "hit count" is a very simple indicator of how people wish to use this amazing tool: the topics that the human mind craves most. Figure 9 shows the hits for a November 2005 search of some interesting words. (*The method of counting web hits has changed since 2005, so all today's counts are smaller.*) Notice the hands-down winners: *Technology* is in first place, followed by *Science*. These two meme-havens together have as many hits as *children, engineering, money, energy*, and *air*, <u>combined</u>.

It's all out there, including aerial imagery of every square meter of the planet, soon every printed word in most of the world's libraries, and today's weather and stock quotes. The way that you *find* anything in that jumble is the magic.

For instance, if you're looking for some arcane piece of information (say, the name of those little mechanical models of the solar system where the planets move about the sun) try typing some of the descriptive words (e.g.: "mechanical model of the solar system") into your search engine, and see how fast you can come up with that obscure word. It's all out there…the search engine brings many websites to your attention, but you can very quickly search through the list to see which one is the likely word you're looking for. Who put that miraculous instant index there? No-one. The indexing of the web is entirely *automatic*: the tool called an inverted B-Tree is what you use to find relevant information.

The inverted B-tree is an astoundingly compact way to index every word (actually, every character) on the internet, storing all the pages on which it's used and where in the page it is. Surprisingly, this indexing database is much smaller than the information it indexes, even though every character on every web page is indexed there. Search engines look for scores of how often each word is used on a page, and score more weight for words that are rarely used (anywhere) as being more specific (and therefore more descriptive) than other words in your list, that are used everywhere. You're searching in context, and there are tricks to weight scores higher when the words you specify are close together on a web page. Good search engines will also know how to stem words, knowing that the words *walk*, *walking*, *walks*, and *walked* are all variants of the word *walks*, and not to ignore them. Really good search engines also know that the word *Walker* is probably not a good match, especially if capitalized, as it is a proper name.

Such context sensitivity is making real strides, driven by the science of computational linguistics pioneered by Ray Kurzweil and others. Consider the following sentence: *John Coates took their coats from the four of them, but put only two of them over there by the fire to heat.* This sentence is a tidy (if somewhat awkward) example of computer recognized context, for it was typed by my computer's speech recognition software. It correctly understood from my rapid dictation (in the same time it would take you to process the spoken sentence) the need for (and spelling of) the proper name "Coates" instead of the plural noun "coats." It also correctly selected the correct versions of the homonyms "for" and "four," "their," "they're" and "there," "two," "to" and "too," "but" and "butt," "by," "buy," and "bye," and the near homonyms "of" and "off." It had to guess at the usually preferred spelling of "John" vs. "Jon." It had twelve chances and sixteen ways to mess up, and didn't. Not bad for software that I got for free.

I have personally helped to develop computational linguistic tools that are very good at suggesting synonyms for your search words. Automatic alternate searches for different (i.e., *correct*) spellings of your keywords are standard in search tools nowadays.

What this means is that the computer is getting smarter (or at least is being coded to act smarter) to assist humans in our imprecise searches. We're not just left on our own in a sea of data. We have powerful tools to assist us, and they are getting more powerful all the time. We can even dispatch web "bots" or "crawlers" to autonomously research compare and filter information from many sources (say, current prices on digital cameras with 5.2 Megapixels), and to prepare a digest of the best options. We've gone from a six-step, six-week multi-intermediary process that occurs at some distance from my office to a one-step, instant process that takes milliseconds, wherever I happen to be. Moreover, the new process is tolerant of my mistakes, and still gets what I want. From a meme's point of view (remember, it's looking for a *mate* out there…) the web is going to promote its evolution several million times faster, and often, *better* than at any prior point in history.

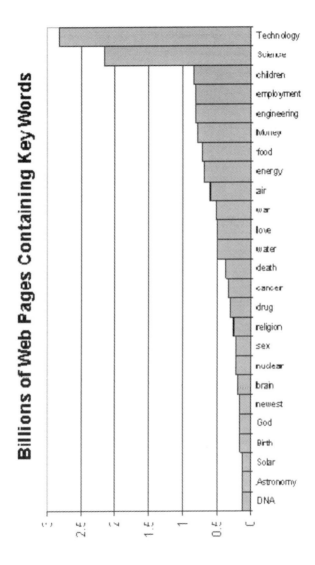

Figure 9. Keyword counts on English web pages.

FACTOR 4: COGNITIVE SCIENCE

4a) A BETTER MAP

We revealed in chapter one that modern cognitive science has mapped out a vast body of knowledge about our own brain's anatomy and chemistry. Recently, functional Magnetic Resonance Imaging (fMRI), Positron Emission Tomography (PET) and other techniques have shown the dynamics of the brain at work: so called six-dimensional (6D) resolution of three spatial coordinates, time, electrical activity, and oxygen consumption rates throughout the entire brain as a function of different mental assignments. We now know the rough function of every few milliliters of the cortex volume. We are aware of at least one unique purpose of each for over one hundred chemicals involved in neural processing, and we understand some of the complex proteomics of how it all works together as a complex system. We know a *lot* about what's in there, even if its complexity seems daunting. This is hundreds or even thousands of times more information than we had a century ago. Remember that one hundred years ago we knew the general function of the tiny Wernicke's area and of Broca's area, the general location of optical processing, the categorization of some neuron types, and that was about it. Only in the last two decades have we developed the non-invasive tools that allow us to study a brain at work in a normal, healthy individual (and of special interest, any abnormally gifted or challenged individual).

Cognitive science has also made great inroads into the overall *integrated* capabilities of the brain, including circadian rhythms, sleep and sleep disorders, optimization (and problems) of attention, problem solving, and memory. Others have studied the effects of various foods, drugs, exercise, and other stimuli on all facets of this precious organ. We have learned from savants the power of *kinesthesia*, or the cross-mapping of certain brain functions into other areas, especially for enhancing creativity.

Armed with this map, we humans have started a wide-ranging quest to improve human thought, from sleep optimization, to diet, to psychopharmacology, to aroma therapy. All of these fronts are designed to help us to learn more, retain information longer, focus better, and to be more creative. To some extent, we're fighting genetics, but that hasn't stopped us in any previous performance measure we've wished to pursue. By diligent study of running techniques, humans have knocked *twenty seconds* (eight percent) off the human record for running a mile distance in fifty-five years. We can surely improve how well we *think* by studying the best practices.

4b) PSYCHOPHARMACOLOGY

In my house I have coffee, tea, several analgesics (ibuprophen, acetaminophen, acetylsalicylic acid, naprosin sodium), sleeping pills, and melatonin. All of these are there to help my brain perform differently than it naturally would, whether the caffeine wakes me up, the analgesics cure a headache (maybe only slightly brain related, but *very* distracting), or the melatonin (or very rarely, a sleeping pill) helps me to sleep. With the exception of tea and coffee, none of these were options to my ancestors only three generations ago. If I were to be diagnosed with the need for Valium, L-Dopa, or Prozac, it would be available.

I take multi-vitamins, and some of them, like folic acid, anthocyanins, beta-carotene, and vitamins C and E are known to be brain-enhancers. (The benefit of such anti-oxidants is clear, given the forty-times-normal usage rate of oxygen in the outer cortex. There's a lot of collateral oxidation going on in the glial cells, leading to many problems. Use of anti-oxidants has shown definitive advantage in reducing the incidence of Alzheimer's and dementia.) Note that despite the hype, the use of ginko bilboa is not a way to better brains, and could dangerously thin your blood. But don't trust me on this…I'm just an amateur. See your doctor before undertaking any new dietary or vitamin program.

Put to good use, these chemicals in my medicine chest and pantry can enhance my mental performance during the day, and improve my rest at night. In the long term, the vitamins may keep my brain free from dementia longer than if I didn't take them, and allow it to be a few percent more productive in my lifetime. Every little bit helps.

4c) CREATIVITY METHODS

Several times in my career I've been challenged to help re-invent my group's processes. Each time, I've had at my disposal an arsenal of facilitation tools (force diagrams, fishbone diagrams, process flow charts, brainstorming, "six hat techniques," etc.) that often have worked well to push us forward. Now that I'm a cognitive scientist, I can see how every one of these tools compensates for certain weaknesses in human brain characteristics, and allows an external representation and/or a combination of minds to rapidly get over such hurdles to solve the problem. A hundred years ago we had to rely on the insight of gifted individuals. Nowadays, we can dissect the techniques that lead to better creativity, teach them to all entry-level personnel, and reinforce the techniques throughout their long, creative careers. We are not only teaching brains better patterns of the physical world, we are teaching them better patterns of how to build new patterns.

COMBINATORIC EXPLOSION

Let's combine the factors and see what we can deduce about our new pace of intellectual growth. First, however, let's make sure that we know the influences that may affect our estimates. We can be wildly optimistic, or dourly pessimistic about almost any factor. In reality, we will probably see a real effect somewhere in between the two extremes.

We've made the assumption that each of these factors is independent of the others: that if all other things remained equal, growth in each area would have a linear effect on our knowledge and understanding. If this is not true, then the numbers don't really multiply. For instance: is global diversity really decoupled from global communications? I believe that it is, based upon first-hand experience in my job duties, where I see as much diversity in my local workforce as I do around the planet. Perhaps they're not fully independent factors, however, and an increase in one may be dependent upon an increase in the other. As another example, will an increase in the archive of the web require greater communications and computing growth, or could we get that gain with the computing infrastructure we have today?

In order to capture these effects, we need to accept that some or all of our estimates cannot be extreme optimistic values.

Other mitigating influences may slow us down as well. If a pandemic illness sweeps the planet, we may suffer a setback that no amount of computing or education will fix in the short term. Similarly, religious extremism and holy wars may grow in the face of the world's technical progress (witness the 9/11 attacks), and this too may damage the globally enriched environment in which scientific brains flourish today.

However, even in the face of the halving of progress that we've encountered during such society-halting events as the bubonic plague and in the Crusades, I believe that the accelerating effects we've explored in the pervious chapter will more than compensate, and still propel us at unprecedented rates. Let's count them up:

1a) We have eleven times the number of educated minds in society now as we did forty years ago. What effect does this have on our intellectual growth? I'll conservatively choose a factor of five in creative output. Perhaps we only pick a factor of three, based upon recently-observed growth in GNP and patent rates, or eleven based on number of degrees, but for now, let's estimate somewhere in-between. It's just a starting point. Soon you'll have a chance to adjust this guess in your own series of estimates.

1b) We have gone from a near ninety-seven percent imbalance in types of minds in the creative core to a 60-40 balance in favor of minorities and women. Minds with predispositions and talents other than those of white males have risen in the creative core nearly twenty-fold in fractional representation. We'll assign a low ten percent (i.e., 1.10) factor for this improvement in diversity, recognizing that diverse opinions lead to more debate, and a possible slower decision process.

    1c) Each adult mind is at work up to seven percent longer than in previous generations. We'll therefore estimate 1.07 as the influence.

    2) There is a critical mass of self-consistent knowledge in which the holes and next steps are more evident, the pattern is clearer, and the rate of solution is faster than at any time in history. We'll be pessimistic and give this a five percent improvement, although this one puzzle factor is the primary cause of the continued once-per-generation doubling of human understanding for the past five hundred years. 1.05.

    3a) We'll be pessimistic and say that this generation will be only ten percent (1.10) more capable now than we are now, based only upon their use of computers as computational tools. Taking the most pessimistic growth rate (Robert's Law, of doubling every twenty-one months) for silicon's impact on a child born today, we see an increase in capability of over four thousand times the current standard before that child turns twenty-one years old. However, experience has shown us that all of that silicon horsepower only slowly solves new classes of problems for us. (E.g.: I still do most of my NASA calculations on a simple spreadsheet, technology over twenty years old.) Thus, 1.10.

3b) Virtually every possession will by 2014 have interactive intelligent features through printed electronics that map it into the parallel universe of information without effort. We'll assign a low importance to this, say 1.05.

3c) Five times the population of the internet is about to join the intellectual debate through the One Laptop Per Child project. An increase of a factor of five in available brains and ideas will be online in just a few years. We could be wildly optimistic that all of these brains will suddenly join the creative core, but in all likelihood most of them will just exist better as individuals, and work to balance the quality of life for the larger masses. Even discounting their poor economic status at the moment, it is credible that out of these five billion new minds, a few percent will make world-value contributions that push our capabilities at the extremes.

Of greatest significance, the Third World lacks much of America's and Europe's infrastructure, and has no economic incentive to hold on to older technologies. They may lead the thrust into distributed, renewable energy systems, completely wireless information architectures, etc.

Communications amongst all people will increase, crossing national, cultural, and economic boundaries. This can only be good. We'll be conservative, and assign a 20% eventual improvement to mankind's total understanding through the impending arrival of truly global data access and communications.

3d) The web now provides virtually every knowable piece of information, and serves up the most relevant pages and pathways in milliseconds, not weeks, to anyone. Bypassing weeks of time and the information bottlenecks of old, every Internet citizen now searches casually for any needed information, usually with instant gratification. We'll conservatively give the web a factor of five in influencing human intellectual growth, but that would be like saying that the printing press was only five times more efficient than hand lettering at propagating human knowledge.

4) We have key enablers to a more productive creative class. We have new self-awareness and new chemical and psychological tools to circumvent our limitations and to enhance our performance. We'll conservatively give this a 2% effect on society. (1.02)

So what does that add up to? Nothing. It multiplies! The cumulative effect, over the next generation, of the factors estimated here are:

$(5)*(1.1)*(1.07)*(1.05)*(1.1)*(1.05)*(1.2)*(5)*(1.02)$

=33 times the current rate, or over sixteen times the previous average rate of acceleration over the next twenty years. Although it is hard to pin down exactly, we see that simple, *very* conservative estimates lead us to believe that instead of doubling our understanding and capabilities every thirty years, we are embarking on a trend that will double perhaps every two years.

You've probably disagreed with one or more of my estimates, and would like to join the debate. Perhaps you've seen how sensitive the number is to the estimate of educated minds, or to the influence of the web, and would like to adjust downwards. Remember that if you get a cumulative factor higher than two, then you have agreed that the rate of acceleration is increasing. If you get a factor greater than four, then you estimate that we will continue at or greater than our current twenty-year doubling rate even in the face of a world pandemic or religious holy war, which traditionally have halved the workforce.

Use the worksheet on the next page to create your *own* estimate of the factors, and enter them at the website:

www.normandyhousepublishers.com/BANG.htm

where you can then see statistics of how others have voted. You see, even in something as mundane as a book, in today's world we can establish a vast feedback system, discussion group, and consensus. The meme of *The Parallel Bang* is out there, and ready to evolve with thousands of other minds joining the debate. Please join in!

	Factor	Estimate
1a	Number of College Degrees	
1b	Diversity of Brains	
1c	Longevity of Brains	
2	Critical Mass in the Puzzle	
3a	Computing Applications	
3b	RFIDs & Smart Printing	
3c	Global Community access	
3d	Web Archive & Search	
4	Cognitive Science & Brain Physiology	
=	**TOTAL PRODUCT** 1a*1b*1c*2*3a*3b*3c*3d*4	

What *else* is affecting the intellectual growth of the planet?

Improving it:

Slowing it:

Other comments:

A SPECIAL NOTE

At this point, I should probably mention the debate that is heating up between religion (pick any one) and the scientific world. As I stated before, religious beliefs are memes that copy themselves from generation to generation. They have been powerful replicators for over five thousand years. It is essential that their central tenants are immutable. Change in religious doctrine comes slowly and painfully. Such memes have brought comfort and peace to humans from times before written history. Religion memes are valuable, and persist because they serve so well.

On the other hand the memes in science evolve, and are *supposed* to be constantly mutating. They are now providing new and farther-reaching answers at an unprecedented rate. We have seen that this rate is growing explosively. This new pace has created an increased intensity in the head-to-head conflict between religious and scientific memes proportional to the number of new scientific findings. They continue to conflict with each other, whether in the form of prohibited books, autopsy bans, legal battles regarding creationist teachings in the classroom, prohibition of women from education, prohibition of embryonic stem cell research, or airliner attacks on the World Trade Center.

In science, one takes *nothing* on faith, and disbelief is encouraged until proof is found. There is no comfort in science memes, except the belief that ultimate truth and some social benefit are somewhat closer as a result of one's pursuits. In any religion, faith is *everything*. The security of knowledge that there is an explanation for the world is the comfort that substitutes for personally seeking the explanation. If the answer is in the mind of God, then it is known, even if we humans do not comprehend it. Demanding proof is actively discouraged, as evidenced in the stories of the Tower of Babel and of Doubting Thomas. Moreover, the set of religious memes in any faith usually includes an optimistic view of life after death, which is a very potent reward for embracing the religious meme set throughout life.

Tension occurs each time science evolves a competing explanation for any particular part of a religion's doctrine. In the recent past, religious and scientific memes could coexist in one brain, largely because the great mysteries that remained unsolved were relegated to God. However, the sheer scope of scientific endeavor in the 21st century will bring a troubling time to our society and to our descendents, as they struggle to balance a testable and ever more complete model of the universe with millennia-old beliefs. Meanwhile, global communication and mobility have brought local religious sects out of isolation and into contact and conflict with each other, putting enormous pressure on each belief system and tradition, even in the absence of science.

More than just the scientific frontier is about to change. If you pray, please pray for peace. If not, please work toward it.

Then you shall know the truth, and the truth shall set you free.
The Bible
-John 8:32

And the Jews say: The Christians do not follow anything (good) and the Christians say: The Jews do not follow anything (good) while they recite the (same) Book. Even thus say those who have no knowledge, like to what they say; so Allah shall judge between them on the day of resurrection in what they differ.
The Qu'ran
-Cow 2.113

The Buddha explained to him…that only when a recluse practiced to become accomplished in morality, concentration, and knowledge: cultivated loving-kindness, and dwelt in the emancipation of mind, and emancipation through knowledge that he would be entitled to be called a samana and brahmana..
-The Tipitaka
Suttanta Pitaka
Mahasihanada Sutta

What thing I truly am I know not clearly: mysterious, fettered in my mind I wander.
The Hindu Rig Veda
Hymn 164 (Visvedevas)
Line 37

CURRENT MEMES IN SCIENCE

PART A: THE VERY BIG UNIVERSE

As we've seen, science memes evolve. Among scientists (i.e., the non-theological side of the debate) the following models have proven to be better than their predecessors in explaining what we see in the universe around us. They are constantly being revised, but they explain a lot of what we see, and the predictions can be tested and verified in many different ways. These memes will propagate until they are supplanted by ideas that are even more self-consistent, but because these are so well supported, I expect that the permutations will be small. In the same way that sharks, crocodiles, cockroaches, and humans are now genetically static, having filled a particular genetic niche to near perfection, I suspect that most of today's memes are nearly at the end of their evolutionary trail, barring some great new scientific or theological revelation.

Here's what we thought six hundred years ago: the Earth is the center of the universe. A vacuum is impossible. The sun, the planets, and the entire zodiac of stars are on spherical crystalline spheres concentric with the Earth. Heaven is outside of the last sphere. Hell is in the center of the Earth, where it is very hot. God put it all there five thousand years ago in only six real days.

THE PARALLEL BANG

Here's what we thought two hundred years ago: the Earth is a rocky planet in an elliptical orbit around our sun, which is made of hydrogen, but does not appear to be burning in the familiar sense. The sun is a star, and is very far from the others, which do not move. A vacuum is possible but abhorrent. Therefore, an invisible, mysterious ether transports light. The Earth is five thousand years old, and the planet is cooling gradually. God put it all there in six equally-spaced chunks of time, whose length is open to interpretation.

Now, here is the stuff we believe today, all of it new during the last human life span, and all of it consistent with observation in numerous branches of physics, astronomy, geology, and chemistry:

Almost all of the volume of the universe is a vacuum. The dispersed matter in the vacuum draws most of our attention, however. Our sun is a small yellow star midway along the "main sequence" of stellar life. The main sequence defines the life of all stars. The death of each star varies, and is predictable based upon stellar mass and the mix of original elements from which the star accretes. Each star is fueled by the fusion of hydrogen through several known steps into heavier and heavier elements.

Stars that are approximately ten times the size of our sun can and do become black holes, whose gravity is so large that even light cannot escape them. A black hole has been proven to exist at the center of our galaxy, and at the centers of almost all others. In late 2005 it was shown experimentally by several cooperating spacecraft that the puzzling gamma ray bursts seen approximately once per day in random regions of the deep universe are in fact the signature of the moment when a star collapses into a black hole.

Our sun is a second-generation small star that formed 4.3 billion years ago out of the remnants of a previous larger star (or cluster of stars) that followed another possible destiny of stars: violent explosion in a supernova. The entire lifetime of our solar system has occurred in the last third of time since the big bang, which created the universe from a point approximately 1.6×10^{-35} meters in size (the "Planck Length," which is the size of a string in the new string theory). The universe is now (after a rapid expansion early on) approximately 156 billion light-years, (or 1.47×10^{27} meters) across, which means that it is sixty two orders of magnitude larger than when it started.

Planet Earth and at least seven other planets[10] formed with our star (Sol), and each planet was bombarded by other solid debris as it accreted from the nebula. Planetary accretion is very common in a star's formation process.

---

[10] Pluto, the 9th planet, is thought to be the largest and closest of a whole family of objects in the Kuiper belt, including the most recently discovered ones: Quaoar, Sedna, and Xena. Pluto may be demoted to planetesimal if the others are not formally recognized as planets. There are presumably thousands of such objects around the sun

Toward the end of our accretion, Earth collided with a particularly large (Mars-sized) proto-planet. That collision split us into our current planet and our moon, tilting us (and the moon's orbit around us) 23.5 degrees to the Earth's orbital plane around the sun. The Earth is the denser core left over from the collision, and still has a thick molten layer below its floating rock crust (and above a solid iron core), kept molten by numerous radioactive elements within. Meanwhile the moon, being smaller and made of lighter ejected material, could not maintain enough internal heat to remain molten, and has therefore solidified.

The Earth's swirling, electrically-conducting molten rock interior creates our magnetic field, and this field flips every few hundred thousand years due to instabilities in the fluid flow. The floating crust is in large chunks called tectonic plates that are pushed around by the convecting magma, leading to earthquakes and volcanoes primarily occurring at plate boundaries. Only 220 million years ago (one twentieth the age of the Earth) the plates were re-clumped together after 400 years of drifting such that a single land mass called Pangea rose above the oceans. The plates have since drifted apart once again to separate the continents.

About 3.8 billion years ago, and after the collision that formed our moon—when most of the matter in the solar system had finally pulled either into the sun itself or into its many planets—the surface of the Earth finally could cool, and liquid water could persist. Microbial life appeared within a few million years of the cooling of Earth's crust, leading to three fascinating scientific possibilities. These possibilities are:

    1) that single "cells" (replicating molecules within lipid bubbles) easily randomly form and are thereafter self-sustaining and prone to evolution in sufficiently wet temperate environments, or

    2) that life in general is highly improbable and that life on Earth was just a real fluke, or

    3) that emergence of life from scratch is improbable, but that it exists in many places in the universe, and is carried via space-borne debris from other life-bearing planets to new ones. If the new planet is suitable, life takes root easily and follows a new evolutionary path (the theory of "Pan-Spermia," first advanced by Sir Frederick Hoyle).

Possibilities one and three have profound implications, and enhance the prospects of intelligent life elsewhere in the universe.

    The Earth still gets hit by residual debris from the accretion of our solar system (asteroids, meteors, and comets) at an average rate of one hundred sixty tons per day, mostly in very fine dust which never penetrates all the way through the atmosphere. The collisions with the largest objects have led to several (if not all) major extinctions on the planet. The Jurassic extinction of the dinosaurs and the Permian (pre-cretaceous) extinction of 99% of all life forms were the two extinctions now strongly believed to be caused by collisions with objects from space. Humans have observed large space debris objects wreaking major destruction on the moon in 1178 and in 2005, and on Jupiter in 1994. The Tunguska event of 1908 is believed to have been a collision with a small body, possibly a very small comet. We now track well over nine hundred potentially life-extinguishing Earth-crossing asteroids, and we find new ones at the rate of about one per month. We expect that the total number of such objects is about two thousand, leading to the observed major extinction of life on our planet every sixty million years or so. We are overdue.[11]

    Large planets are evident around half the nearby stars that we observe, and smaller rocky ones like ours may exist around most or even all of them. Once we build instruments sensitive enough to observe them, this question should be resolved. We are designing and building the necessary instruments now. We observe the formation of new stars out of great nebulae of dust and gas, and we see the accretion disks of materials that we believe to be the precursors of planets around such stars. Meanwhile, we have discovered several new bodies outside the orbit of Pluto, and now believe that there are thousands of such things farther away from our sun. The residual non-accreted matter around our sun is in

---

[11] As Arthur C. Clarke once observed, "The dinosaurs died because they didn't have an effective space program."

three major regions: the asteroid belt (mostly between Mars and Jupiter, but extending well inside Earth's orbit), the Kuiper belt (beyond Neptune), and the Oort cloud, a spherical shell well beyond the other matter.

Our star is in the outer third of a spiral arm of a one hundred thousand light-year wide galaxy, which is one among one hundred billion galaxies which are themselves distributed in space on the expanding walls of a foam-like series of spherical thin-walled bubbles. The center of our galaxy in is the direction of the constellation Sagittarius, but the center of the universe itself is not known, due to the peculiar mathematics associated with a uniformly expanding volume such as our universe. In this geometry, any point where the observer happens to be appears to be the exact center of expansion. Wherever the center is, all matter has been propagating outwards from this point and from a single explosive moment 13.7 billion years ago. (We know the time by measuring the rate of outward expansion of objects near to us, not by observing the true edge of the universe, which cannot be done by electromagnetic waves such as light.) The foam-like structure of the universe is now thought to be a quantum mechanical artifact from the first quintillionth of a second of its existence. The quantum foam left a mixture of voids in the infinitesimally small universe, and after the first instant of expansion, the scales were such that the voids were permanent. That's today's thought, at least.

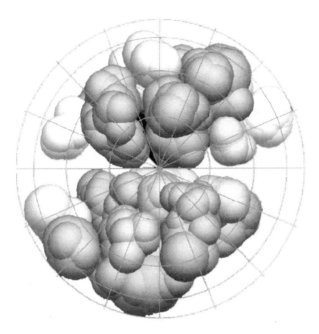

Figure 10. The large scale structure of the visible universe. Each of the spheres is essentially empty, and the average size is 50 MegaParsecs across. Most observed matter is on the walls of these spheres, and a galaxy is smaller than a pixel in this scale. (Our Milky Way galaxy is less than $1/1000^{th}$ of the size of any of the spherical voids). One parsec is thirty-one trillion kilometers, so each sphere is about 1.5 sextillion (or 1,500,000,000,000,000,000,000) kilometers across. The horizontal seam in the middle is likely also filled with such spherical shells of matter. The gap in data is caused by the many stars in our own Milky Way galactic plane blocking our view. Graphic by Shai Ayal, appearing in H. El-Ad, T. Piran, and L. N. da Costa, *Mon. Not. R. Astron. Soc.* **287**, 790–798 (1997). Used by permission.

PART B: OUR UNDERSTANDING OF LIFE

We've known since the fifth century B.C.[12] that matter was made of atoms. Later we found that atoms could be combined to make molecules, and realized only lately that some very complicated molecules called proteins exist in our bodies. We believed that unknown chemical signals called "genes" were shared in offspring between male and female parents of all plant and animal species, and we had started to gather statistics of which traits were dominant and which were recessive.

Today we believe that every protein is built of precise sequences of only twenty basic building blocks called amino acids, and that each protein is built by elaborate (but explainable) automatic molecular patterning processes. In the last four decades we've learned that a protein's assembly is conducted at strands of Ribonucleic acid (RNA). RNA is itself patterned—as an offset printing press produces printed copies from a master—by Deoxyribonucleic acid (DNA). DNA is the complex molecule that stores genetic information in the nucleus (and mitochondria) of each cell. DNA is a double coil of phosphates and sugars holding a spiral staircase of four bases:

---

[12] Two philosophers (Leucippus of Miletus and Democritus of Abdera) are credited with developing the premise of atoms and with making early observations of Brownian motion and other indicators that led to support of the theory of such unseen smallest of the small particles.

Adenine, Cytosine, Guanine, and Thymine. RNA is single-stranded, and in it, there is no Thymine. Uracil substitutes for Thymine in every RNA function. RNA transcribes the code in DNA to become a worksite where proteins are built. The four interchangeable building block "rungs" of DNA (and their paired versions on RNA) form a fundamental *genetic* code We know the code of how every three rungs of the ladder correspond to each of the twenty amino acids.[13]

We have transcribed a human- and computer-readable rendition of the genetic coding at every rung of the average human DNA, and the DNA of many other plant and animal species. We are currently learning to interpret these "books" that we have transcribed from the chemical chain. There are over three billion pieces of the DNA code, but we differ from each other by less than one in a thousand of these rungs, more properly called *alleles*. Even so, this limited differentiation

---

[13] Taking account of the ways that each amino acid bends a protein differently into its complicated shape, researchers have proven that the chances of developing a better code through some other encoding scheme (out of the nearly infinite available options) is less than one in a billion, if the goal is to build a protein exactly like, or extremely close to the originally intended protein. (As Dawkins predicted, one necessary survival trait of a replicator to is to make copies of itself as accurately as possible. One of the most accurate of all possible copying techniques is apparently at work in Earth-based life.
(See http://genomebiology.com/2001/2/11/research/0049)

gives every person over three million ways to be different from anyone else. We're learning which one allele in a thousand makes each difference in our bodies, and which are pretty much standard. If DNA rungs were printed words, the difference between any two people would be one word in every three pages of a book, and the book would be ten million pages long. Everyone has some variant transcription of the average book (some by inheritance, some by mutation), but not every difference leads to any significant change.

Although the entire genetic code is carried in every cell of any plant or animal, only certain attributes are turned on in any given cell. The activation or deactivation of select parts of the code to differentiate say, liver cells from brain cells, is thought to be controlled by a combination of the chromatin (attached to the phosphate sugar backbone of the DNA spiral), pieces of "blocking" RNA strands that deactivate some codons, and free-floating "messenger proteins" that are manufactured in and preserve the specialization of each type of cell. The study of the chemical pathways that control cellular function is the field of *proteomics*. Proteomics, because of the near infinite combination of possible chemical chain reactions, is a rich field of study that will be strongly accelerated by computer modeling and simulation during the Parallel Bang.

The theoretical understanding of biochemistry is struggling to keep pace with horticulture and animal husbandry, where our understanding of desirable traits has allowed humans to create thousands of useful variants.[14]

PART C: OUR UNDERSTANDING OF THE INFINITESIMAL

Moving downward from the molecular world of biochemistry, we knew a century ago that atoms contained electrons around charged nuclei. Less than a hundred years ago Ernst Rutherford found the neutron, and also calculated its size. Since 1964, we've confirmed that a whole "zoo" of even smaller subatomic particles exist, currently nineteen in number, and more are predicted at very high energies.[15] Enormous,

---

[14] Floyd Zaiger of Modesto, California has personally created over 200 varieties of fruit, and his white peaches, not commercially available in 1990, now constitute 22% of all the peaches produced in the United States. Meanwhile, milk cows now produce twice the milk that they did two human generations ago.

[15] The known particles are in three families of physical objects and one family of "force particles." (Remember that light has particle-like properties, including quantized energy levels, which is best modeled by the existence of real particles, which we call photons. Similarly, all other forces of nature have quantized behavior, and are assigned mathematical particle representations. The force particles are

Photons (electromagnetic force)

Gravitons (gravitational force),

Gluons (nuclear strong force, holds Neutrons and Protons together in a nucleus), and

powerful instruments are being built (particularly at CERN in Switzerland) to confirm if these predicted other particles are real. Neutrons and Protons are combinations only of Up-quarks and Down-quarks, and all known stable matter in our part of the universe is made only of electrons, neutrons, and protons. All Family 1 particles have equivalent anti-particles, which if brought in contact with their counterparts, completely annihilate each other resulting in pure energy. Antiparticles are rare in our part of the universe, but may be abundant elsewhere. Massless particles like photons and gravitons carry information about the forces of nature, of which we know four: electromagnetism, gravity, the nuclear strong force, and the nuclear weak force.

The "quantum" effects predicted by the world of subatomic particles are used in such modern technologies as the microwave oven in your kitchen, the laser in your CD player, the semiconductors and LEDs in your computer, as

---

Weak Gauge Bosons (two types, W and Z, each of which hold combinations of the primary subatomic particles like quarks together in assemblies under the nuclear weak force). The families of physical particles are:

Family 1: Electron, Electron-neutrino, Up-quark, and Down-quark.

Family 2: Muon, Muon-neutrino, Charm-quark, and Strange-quark.

Family 3: Tau particle, Tau neutrino, Top-quark, and Bottom-quark.

well as more esoteric technologies such as Cesium clocks, superconducting wires, and MRI Scanners.

Now for the kicker…we now believe that *all* of these particles and forces are manifestations of the *same thing*: tiny loops of energy called *strings* that vibrate in different ways. M Theory, or the "Theory of Everything," is centered on some very difficult math that may just explain everything we see in the universe.

SIXTY-TWO POWERS OF TEN

We have a few pieces left to fill in, but the puzzle is coming together fast. Our model of subatomic physics links very well to our understanding of forces and matter in all larger scales. That the subatomic particles build atoms and molecules is consistent with what we see. Our models of how the molecules interact, based upon these underlying forces and particles, is consistent. It also provides thoroughly consistent insight into how biological processes work. Our understanding of even larger forces and objects explains our weather, our continents and all natural phenomena. Our understanding of how astronomically-large blocks of matter coalesce by gravity, convert into energy by fusion, and create new atoms is consistent with the models too. The universe as we know it fits remarkably well in one cohesive set of models. Our models of geology, nuclear physics, optics, Newtonian physics, relativity, astronomy, biology, thermodynamics, mathematics, etc are all proving to be mutually supportive, indicating that we are close to comprehending the integrated, comprehensive truth of the universe.

At least, we are close to understanding the rules (or even, *rule*) of how it works. The astounding complexity that propagates from our simple rules, such as turbulence emerging from a steady flow, or proteomics, can still dazzle us for centuries, but at least we'll understand the root of what is happening at all scales of the physical world. It has taken 13.7 billion years for the universe to grow to its current size. The parallel universe of human intellect has coalesced out of these new models at an increasing rate, reaching critical mass of total self-consistency in just the past few decades: one billionth of the age of the physical universe. In the parallel universe of human understanding, we can span the real universe's size and time in milliseconds, and comprehend and predict things septillions of times larger than ourselves, and octillions of times smaller than a single synapse. All sixty-two orders of magnitude from the smallest to the largest scales now fit in a cohesive, unifying set of rules in our parallel universe, just as they fit in the physical universe. And as we approach perfect understanding of the laws of the physical universe, we can transfer that meme to the next generation for them to improve upon.

WHAT STILL DOESN'T FIT

Among the things that don't fit, we know that the laws of the very small things and very big things do break down, but we've traced that problem to the fact that we've used math that *approximates* reality to assume that things can have absolutely zero size. The new theory, formerly called string theory, but now called M-theory, is close to resolving this problem, and can actually predict why we have the particles and forces we observe in the universe from a single (for now, very complex) equation. One has to start by believing that vibrating strings of energy exist in small loops less than ten millionths of one millionth of one millionth of one millionth of one millionth of one millionth of a meter across, and then it all starts to fall into place. One has to assume the existence of such things beyond the ability of our senses (and even of our instruments) to detect. This is the unique terrain of the mental universe—all other animals can imagine and react to only what they can see, hear, smell, taste, and feel.

There appears to be some matter missing from the physical universe in our mental model of it—so called "dark matter"—because the laws of gravity and the expansion rate of the observed matter don't quite line up. The dark matter could be in free-flying particles, black holes, dust clouds in free space, or even in the centers of those unexplained spherical-wall distributions of galaxies. It's just clear that there's not enough matter in all the self-illuminated stars and galaxies that we can see to cause the rates of expansion that we observe by gravity alone. However, I'm not worried. The most illuminated grey matter in history is working on it.

WHAT FITS, BUT IS JUST PLAIN HARD

In biology and natural physics, we have unlocked the simple rules, but we have the coming problems of chaos, complexity, and emergent behaviors to unravel. The science of proteomics needs to unravel how one hundred fifty thousand proteins interact with each other and with the cellular machinery to make each known biological function, and we need to finish deriving the meaning of each of the three billion alleles on our DNA.

Factoring of large numbers is still hard for our current computers and for the organization of our minds. The number of possible outcomes from even a few simple truths can become unfathomably complex in a short time.

THE COMPLEX PROBLEMS AHEAD

Emergence is the next great frontier in human understanding, once we reach the bottom of all knowledge with the theory of everything. One of the simplest and most beautiful examples of emergent behavior is the Mandelbrot set, named for Benoit Mandelbrot. It is the result of iterating a very simple function ($f(z)=z^2-u$) over and over again in a small region of space, where z and u are complex numbers.[16] Just as the Mandelbrot set and other fractal patterns emerge from incredibly simple functions iterated over and over, such rich

---

[16] A complex number has a real coordinate and an imaginary coordinate, (say, for example, the length direction is real but the width direction is imaginary. Only the human mind, of all animal brains, could conceive of an imaginary number, but they are quite valuable, and explain most modern physics. Perhaps the real world is rooted in imaginary numbers, and our brains, being the only ones that can imagine such things, are the only ones that will *ever* explore that universe.

natural phenomena as intricate ant colonies and the workings of the human mind are far more than the mere sum of their myriad repeated simple parts. The final result is far more wondrous, surprising, and rich.

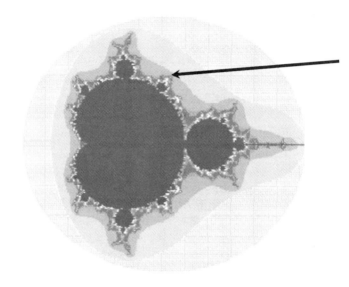

Figure 11. The Mandelbrot Set in the region:
 x from -1.0 to 2.0; y from -1.5 to 1.5 in the complex (imaginary) axis. The dark regions are where the real part of the answer has already converged to a number less than 1. Shaded regions have not converged to either 1 or infinity because not enough iterations have occurred at the point that the plot was generated, or because there is not enough resolution to show the fine detail. (Arrow marks the region magnified in the next figure.)

Figure 12. A zoomed view of one region of the Mandelbrot Set:

> x from 0.54360269 to 0.58063973
> y from 0.62024691 to 0.66514029.

Such intricate fractal patterns from even a simple equation iterated a large number of times can be observed at any level of magnification, even millions of times zoomed from here.

WAS THE PARALLEL BANG INEVITABLE?

There seems to be growing evidence that complexification is the natural order of things although it does seem counterintuitive from everyone's favorite law of thermodynamics (the tendency of everything, on average, to run to a state of disorder). You would think that complex things would tend to break apart, rather than to build into more and more complex structures. On the whole, you'd be right, but there are marvelous intricate things that go on when you stir things up a little.

To understand why this trend towards complexity is natural in *some* areas of the universe, you have to begin with the idea that the natural trend towards disorder is only the norm when energy is flowing from one place to another in a uniform direction. Whenever it doesn't do so in a smooth, uniform way, we have the possibility of interesting new things happening. Typically, of course, everything we see in nature is chaotic, turbulent, and in a state of flux, because we have a lot of energy flowing across our planet in *cycles* that last from an instant (say, lightning) to hours (day/night temperature cycle), to months (winter, summer, spring and fall), to years or decades (the eleven-year cycle of sunspots first observed by Galileo, which is now known to correlate with huge energy fluxes from the sun), and even to millennia, as seen in geologic records..

The flow of energy through a system with any sort of instability or cycle is governed by a field called *non-equilibrium thermodynamics*, and in this field, the tendency for everything to run towards disorder (the second law) is actively battled by the first law, which says that heat (energy) can be turned into *work*, which, when you get right down to the essence of things, is the restoration or creation of *order*. Start with a pile of wood, apply some work, and you get a house. Fail to periodically add a little more work, and it all runs to disorder. Apply a lot of work, and you get a major home improvement.

Basically, every cycle of energy flow has the potential to be a kind of engine. The ways in which the energies move in and out of a system all have names (Carnot Cycle, Otto Cycle, Diesel Cycle, etc). Each of these cycles is known to do work, i.e., to create order in some areas at the expense of further disorder elsewhere. Like the Mandelbrot set, we can imagine that complexity can build to new levels if we allow a very simple function to operate over and over and over again.

We think that our little corner of the universe is a pretty average place, but look at all the swirling cycles of energy going on. The sun's energy reacts with our spinning planet to create patterns of hurricanes, water currents, dew, updrafts and yearly seasons. Waves churn along the shores, aided by the tug of the moon which creates our tides. Sunlight flows in by day, and heat radiates out by night. Hurricanes form in summer and periodically tear up the coastline. Meanwhile, the conductive core of the planet swirls and ultimately stratifies, uplifts, and consumes enormous sheets of different materials.

We expect that these stable cycles are mimicked in many ways in and around most stars in the universe. There are instabilities that occur at scales far greater than our sun, and far smaller than ourselves, all of which give rise to stable complexities such as we observe in snowflakes, ocean currents and eddies, continental drift, volcanoes, and in simple mathematical functions.

Some of these instabilities allow chemical reactions to be made and unmade as energy flows to and from. You might expect that on average for a given mix of chemicals you would get the same statistical mix of products every day. But consider if some of these products, instead of needing to be assembled by accident of the component parts, act like a catalyst within this flow of energy to help make more copies of themselves. These molecules, however simple, will disproportionately swing the use of resources towards more copies of these replicators, and less copies of the non replicating, or inert, chemicals.

Thus non-equilibrium thermodynamics leads us to the idea that complexity can grow, and the idea of replicating versus non replicating complex molecules strongly favors rapid emergence of anything that can make (or lead to the making of) a copy of itself in this environment.

Amongst such systems, consider the thermodynamics of evolution. If an organism accidentally adapts to perform in more than one niche, say, to be able to consume more than one type of resource, or to use an additional sense, it is more likely to survive and prosper in the environment where its peers are limited to the old single way of doing things. So, in the short term, the new variant will surge ahead. Of course, doing more than one function takes more energy and more resources in general, and so the next big step in evolution is that from amongst this population of new multi functional chemical paths (or organisms), any that happened to stumble upon a more efficient way of doing the new function, or of using the new resources, will start to prosper at the expense of the first generation of multi functional copies. You start by getting more complex in order to expand your niche, and then you get more efficient at doing what you're doing in order to excel there. In the case of humans, we have built up a complex brain to work in the niche of nature that is governed by information alone. Now, as we evolve in our intellectual world, we do it by becoming more and more efficient in how we gather, digest, and act on information.

Reciprocating flows of heat, such as are natural in planets around suns, create useful work, and work is needed to create patterns out of chaos. Patterns that copy themselves grow preferentially over those that are just random occurrences. Patterns whose copies can slightly mutate can build bigger advantages over those that don't. Those patterning systems that can adapt to cover more resources and do more functions will outperform those that don't. At the moment, it appears that life is one such replicating pattern that now supports a second type—memes—which after 13.7 billion years are finally exploding into a new universe of their own.

THE SIXTH WAVE?

We explored the first five waves of growth in the information universe in Chapter One. We saw that each wave came approximately one hundredfold faster on the heels of its predecessor than the previous cycle. As memes move out into the *silicon* universe of data, will they outstrip the human mind? Are there tools and methods that may lead to yet *another* explosive growth? It's a sobering thought, but I believe that there will soon be such tools, and a sixth wave will launch.

After all, the facts and equations and programs in the silicon world are just patterned information, just as it is only patterns that propagate in our neurons and our genes. If computer viruses can replicate and propagate in the silicon world, why can't more useful patterns be set to work?

We already send our "bots" and search engines out to distill and refine information from our accumulated knowledge, i.e., to create more order out of the existing patterns. The computer never needs to sleep, and has capability far beyond our ability to direct its resources. Therefore, background computer processes script themselves to do useful work while we're not otherwise directing the system (such as virus scans, or searching for signs of intelligent life in the radio spectrum records).

Genetic algorithms can already build better codes by mutating previously successful code fragments, and codes have evolved to be very good at deducing meaning from our error-prone dictates. Replicating patterns that can mutate and selectively evolve are thus already in the silicon environment. How big a jump is it to believe that soon such bots and genetic algorithms will be self-directed, looking for trends and synthesizing facts of their own from the global database? We will of course be in control of the initial genesis of such programs, but like killer bees, computer viruses, and Max Headroom, I believe that it is all but inevitable that some of these bots may take on a permanent life of their own in yet another parallel universe of understanding, out in our silicon network. Ray Kurzweil has termed this moment "the singularity," and he believes that it is very near. I agree that it is very close.

Maybe these meme-generating ethereal entities will find epidemiological trends, such as increased cancer incidence in and around previously unrecognized sources. Perhaps they'll choose to look for (evolve) a theory that matches the observed sunspot cycle, or a refined theory of Earth's weather. Maybe they'll beat the stock market.

If they do, we'll probably reward them by letting them have the power and silicon they need to grow. And if they do grow, then our *Parallel Bang* will seem tame by comparison.

Plus ça change, plus ça reste le même chose.

-Voltaire

INDEX

Access, 69, 101, 116
Adrenaline, 49, 155
Advertising, 96
Agriculture, 9, 26, 34, 37, 76, 77, 81
Aircraft, 103, 156
Alvarez, Dr. Walter, 65, 88
Asteroids, 65, 88, 126, 127
Astrobiology, 12
Atoms, 11, 17, 28, 89, 130, 133, 136
Autism, 52, 57
Automobiles, 13, 14, 16, 96
Brain, human, 10, 13, 16, 17, 18, 27, 29, 30, 32, 33, 46, 47, 49, 50, 51, 52, 53, 55, 60, 61, 63, 64, 65, 67, 69, 73, 74, 75, 76, 77, 78, 81, 83, 85, 91, 93, 101, 103, 110, 111, 113, 114, 115, 117, 120, 121, 139, 140, 146, 154
Brownlee, Dr. Donald, 30
Cambridge University, 74
Chemistry, 26, 49, 88, 108, 123
Chemistry, 26, 49, 88, 108, 123
Children, 42, 68, 101, 104
China, 23, 154
Clarke, Dr. Arthur C, 25, 127
Computers, 10, 11, 13, 14, 36, 37, 62, 93, 94, 95, 98, 99, 100, 102, 105, 106, 114, 131, 132, 134, 139, 146, 147
Culture, 22, 24, 25, 35, 57, 70, 73, 83, 115
Davies, Dr. Paul, 12
Dawkins, Dr. Richard, 64, 65, 66, 131
Delbrük, Dr Max, 88, 89
Democritus of Abdera, 130
Diamond, Dr. Jared, 25, 33
Diversity, 52, 55
DNA, 17, 32, 64, 88, 89, 130, 131, 132, 139
Dopamine, 49, 54
Education, 39, 61, 79, 81, 82, 95, 100, 113, 119, 153, 154, 159
Feynman, Dr. Richard, 16
Genes, 27, 33, 37, 53, 58, 64, 65, 66, 130, 146
Germany, 80, 102
Great Leap Forward, 32
Harvard, 51, 53, 61
Hawking, Dr. Stephen, 16
HDTV (High Definition Television), 39
Health, 77
Highways, 16
Homo Erectus, 31, 92
Homo Sapiens, 26
Houston, TX, 2, 4, 159, 160, 161
Hydrogen, 123
India, 80, 82
Internet, 101, 116
Japan, 82
Kurzweil, Dr. Ray, 9, 105
Languages and English, 103, 107
Laws, 10, 11, 36, 94, 97, 100, 114
Left-handedness, 46, 54, 55, 56, 61
Leucippus of Miletus, 130
Life expectancy, 9, 11, 85

Lord Kelvin, 16
Mandelbrot Set, Dr. Benoit Mandelbrot, 139, 140, 141, 143
Mars, 54, 125, 127
Medicine, 9, 25, 110
Melatonin, 49, 110
Memes, 53, 65, 66, 67, 73, 83, 85, 92, 93, 101, 104, 106, 117, 119, 120, 122, 137, 146
Money & Trade, 77, 119
MRI (Magnetic Resonance Imaging), 135
Negroponte, Dr. Nicholas, 100
Nobel Prize, 88, 89
OLPC (One Laptop Per Child) Foundation, 100, 102
Oxytocin, 49
Patents, 39, 60, 71, 75, 113
PET (positron Emission tomography) scans, 48
Population, 9, 22, 34, 35, 55, 60, 68, 74, 75, 76, 77, 80, 81, 82, 85, 86, 87, 100, 115, 145
Proteins, 42, 89, 130, 131, 132, 139
Proteomics, 132
Quaoar, 124
Religions, 9, 35, 66, 68, 71, 113, 117, 119, 120

Renaissance, 23, 24, 34, 37, 86, 92, 155
RFID (Radio Frequency Identifier tags, 15
Roberts' Law, 97, 100
Russia, 35, 69, 158
Sagan, Dr. Carl, 30, 158
Sedna, 63, 124
Senses and sensory input, 44, 50, 138
Smalley, Dr. Richard, 16
Solar Energy, 12, 27, 62, 64, 98, 104, 124, 125, 126
Stock, Dr. Gregory, 62
Summers, Dr. Lawrence, 53, 61
Supercomputers, 95
Superconductors, 135
Technology, 9, 12, 23, 25, 31, 34, 36, 37, 77, 78, 80, 82, 92, 97, 99, 115, 153, 158
Third World, 115
Toffler, Alvin, 9
Transportation & Travel, 9, 25, 26
Universe, Parallel, 2, 4, 98, 117, 148, 158, 160, 161
Valium, 49, 110
Ward, Dr. P.D., 30
Women, 79, 81, 82
World Wide Web, 2, 4, 37, 51, 117, 154, 159, 161
Xerox Corporation, 95, 158

Book a Live Event!
History in the Making

From the schoolhouse out to the global arena and into the cosmos, *History in the Making* prepares today's professionals for the astounding journey ahead. Your staff will be energized and motivated to develop the full potential of every student in your district and every employee of your creative staff as key participants in this pivotal period of human history.

With rave reviews from around the world, Jack Bacon's *History in the Making* program is an educational and inspiring event that is especially well-suited as a kickoff to the corporate or academic year. It's not about history that has been. It's about the history that is yet to be, made richer by those with a complete and diversified education, interacting with tomorrow's technology. Beginning with participatory exercises that demonstrate how the human brain acquires and stores information, Jack explores the educational process and demonstrates the potential of every educated individual in each new generation. He shows how each generation builds upon the knowledge of the ages. By adding to that knowledge, each generation evolves its society, makes its mark in history, and advances the human species. Jack ties all his teachings together with the evidence and the rationale of how and why the pace of human intellectual development has begun to accelerate to unprecedented rates in the current generation. Such development has gone from an historical doubling time of approximately once per generation to a modern pace many times faster. *History in the Making* ends with a thrilling tour of some of the technologies, trends, and challenges that will form the foundation of the world that your staff will shape. Following the lecture, Dr. Bacon eagerly consults with groups of your specialists, fostering innovative thoughts on all fronts.

It's the perfect start to your year. Full of hope, help, and vision, *History in the Making* elevates the importance of education and celebrates motivated contributors in all subjects. It explores how we learn, how we create new knowledge, and how your staff has the unbounded potential to advance all of humankind. It celebrates the individual protégé and the individual mentor. Most importantly, it brings satisfying and fulfilling perspective and pride to those who accept the unprecedented challenge and importance of teaching and reaching young minds in today's world.

History in the Making has already dazzled audiences across England, Scotland, Wales, Canada, Israel, China, Australia, New Zealand, and dozens of U.S. states. Portions of this presentation have been incorporated by the Texas Education Association in the motivational video *Countdown to Your Future*. Professional groups in dozens of companies have raved about this event. Now you can have a live presentation for your staff and for the teachers of your local school district! They'll be talking about it for years.

Technologist, futurist, historian, educator, author, cognitive scientist, and worldwide speaker Dr. Jack Bacon is scheduling speeches now. After sixteen years, thirty-one countries, and over a thousand appearances, he's packaged his very best work in a comprehensive, uplifting foundation especially for you. Don't miss this chance! Past clients have been so thrilled with Dr. Bacon that we guarantee your audience's satisfaction.

TOLL FREE: 1-866-447-4622
INTERNATIONAL 1-281-814-8665
www.DrJackBacon.com

Order *My Grandfathers' Clock*!

My Grandfathers' Clock immerses you in every era, with real characters you care about, living lives like yours. You'll be a part of history as real people lived it: not as kings and explorers, but as clerks and soldiers and merchants, students, teachers, lawyers and tradesmen. These are the people who made and lived the grand saga. These are your ancestors, and you can find your heritage through them.

You will feel the adrenaline of medieval battle and the oppression of feudal life. Your frustration will build to a traitor's passion and a crusader's zeal. You will feel empowerment as you contemplate the changes that the *Magna Carta* will make to your life. You will search for truth in the early universities, only to face terror as the Black Death enters your home.

Your spirit will soar in the art of the Renaissance, and your senses will be challenged by the discoveries of the New World. You will question your faith, as you walk the thin line between treason and heresy in Tudor England's *Via Media*. You will agonize, and then resolve to tear your family from your ancient home, to build a new, purer life in America. You'll face mortality in Massachusetts, with only herbal medicines between you and the grave. You'll prepare for revolution, and then make ends meet in the economic ruin that follows. You'll exhaust yourself opening the American west, and rage against slavery in the pulpit, as you await the latest news from Fort Sumter. You'll open the world's eyes to the wonders of communications, the airplane, and finally, of space.

My Grandfathers' Clock is a story so broad, so moving, and so powerful that no one person could invent it. It is spectacular because it is real. Twenty-eight generations of one family bring history to life, each father contributing a single day to history, so that his son can carry on. Part of your own family history is likely in this book: over a billion people have branched from this ancient lineage. Celebrate the bond of fathers and sons through a millennium of the Anglo-American experience!

You'll find that there's more to *My Grandfathers' Clock* than a great story:

You'll learn
You'll laugh
You'll be inspired
...You'll be blown away.

Didn't "get" history? Get this book!
Got a father? Get this book for him! (But read it first!)

ABOUT THE AUTHOR

Jack Bacon has often been called "A New Carl Sagan." He is an internationally-known motivational speaker, an emeritus distinguished lecturer of the American Institute of Aeronautics and Astronautics (AIAA), and one of the most requested speakers in the country for topics concerning the technology and its impact on society. He is a noted futurist and a technological historian, and has written three popular books entitled *My Grandfathers' Clock*, *My Stepdaughter's Watch*, and *The Parallel Bang*. He inspires over one hundred twenty audiences per year. His lectures have captivated tens of thousands of all ages in thirty-one countries on six continents, and he has appeared on numerous radio and television broadcasts.

A graduate of Caltech (B.S. '76) and the University of Rochester (Ph.D. '84) his extensive career includes roles in the development of many cutting edge technologies, including controlled thermonuclear fusion, the development of the electronic office, factory automation, and the globalization of business. He pioneered the deployment of several artificial intelligence systems, learning his craft at the famed Xerox Palo Alto Research Center. Since 1990, he has been a key player in the development of the most difficult and global engineering project in history—the International Space Station.

He was the United States' lead systems integrator of the *Zarya*—the jointly-built spacecraft that forms the central bridge and adapter between all U.S. and Russian technologies on the Space Station. This landmark in technological history was built in Moscow by American and Russian engineers and launched from the Baikonur Cosmodrome in November 1998.

Jack is a fellow of the Explorer's Club, a member of the National Speakers' Association, The International Federation of Professional Speakers, Engineers Without Borders, the AIAA, and he was a founding member of the board of directors of the Science National Honor Society (see www.ScienceNHS.org). He routinely advises numerous academic programs and institutions, and he is a champion of education throughout the world.

Photo for NASA by Robert Markowitz

Do you want to purchase this or other **Normandy House** publications?
Call Toll Free: 1-(866) 447-4622
International: 1-(281) 814-8665

Visit us on the web at
www.normandyhousepublishers.com
Or, mail us at:

Normandy House Publishers
P.O. Box 59-1066
Houston, Texas 77259-1066

Want to make some history of your own?

Award-winning speaker Jack Bacon, author of ***My Grandfathers' Clock***, ***My Stepdaughter's Watch,*** and ***The Parallel Bang*** is available to provide any of several *History in the Making* seminars or speeches to your organization. A veteran of dozens of international television and radio broadcasts and hundreds of dazzling presentations in thirty-one countries, Jack Bacon has entertained the world, and can make your next event an historic moment to remember.

Contact:
New Epoch Productions
P.O. Box 59-1066
Houston, Texas 77259-1066
On the web:
www.DrJackBacon.com
Or Call <u>toll free</u> **1-(866) 447-4622**   to arrange it.